Agricultural Robots

Mechanisms and Practice

Edited by

Naoshi Kondo
Mitsuji Monta

and

Noboru Noguchi

Kyoto University Press

First published in Japanese by Corona Publishing Co., Ltd. as

農業ロボット（II）— 機構と事例 —
近藤　直・門田充司・野口　伸　編著
コロナ社　刊

Agri-Robot (II) — Mechanisms and Practice —
Edited by Naoshi Kondo, Mitsuji Monta & Noboru Noguchi

This English edition published in 2011 jointly by:

Kyoto University Press
69 Yoshida Konoe-cho
Sakyo-ku, Kyoto 606-8315
Japan
Telephone: +81-75-761-6182
Fax: +81-75-761-6190
Email: sales@kyoto-up.or.jp
Web: http://www.kyoto-up.or.jp

Trans Pacific Press
PO Box 164, Balwyn North, Melbourne
Victoria 3104, Australia
Telephone: +61-3-9859-1112
Fax: +61-3-9859-4110
Email: tpp.mail@gmail.com
Web: http://www.transpacificpress.com

Translated by Minako Sato.

Edited by Karl Smith for Trans Pacific Press.

CD-ROM prepared by KWIX Co. Ltd.

Distributors

Australia and New Zealand
DA Information Services/Central Book Services
648 Whitehorse Road
Mitcham, Victoria 3132
Australia
Telephone: +61-3-9210-7777
Fax: + 61-3-9210-7788
Email: books@dadirect.com
Web: www.dadirect.com

USA and Canada
International Specialized Book Services (ISBS)
920 NE 58th Avenue, Suite 300
Portland, Oregon 97213-3786
USA
Telephone: (800) 944-6190
Fax: (503) 280-8832
Email: orders@isbs.com
Web: http://www.isbs.com

Asia and the Pacific
Kinokuniya Company Ltd.

Head office:
38-1 Sakuragaoka 5-chome
Setagaya-ku, Tokyo 156-8691
Japan
Telephone: +81-3-3439-0161
Fax: +81-3-3439-0839
Email: bkimp@kinokuniya.co.jp
Web: www.kinokuniya.co.jp

Asia-Pacific office:
Kinokuniya Book Stores of Singapore Pte., Ltd.
391B Orchard Road #13-06/07/08
Ngee Ann City Tower B
Singapore 238874
Telephone: +65-6276-5558
Fax: +65-6276-5570
Email: SSO@kinokuniya.co.jp

The translation and publication of this book was supported by a Grant-in-Aid for Publication of Scientific Research Results, provided by the Japan Society for the Promotion of Science (JSPS), to which we express our sincere appreciation.

ISBN   978-1-920901-83-7

Contents

How to use this book and the attached CD-ROM

Figure 1 shows a sample of the layout. Under each section title, general descriptions of the project development organizations and the year of project completion are provided. In sections where (movie) or (movies) is specified on the right of the title, the attached CD-ROM contains the relevant motion video(s). The CD-ROM also contains the pertinent figures and photos, mostly in color.

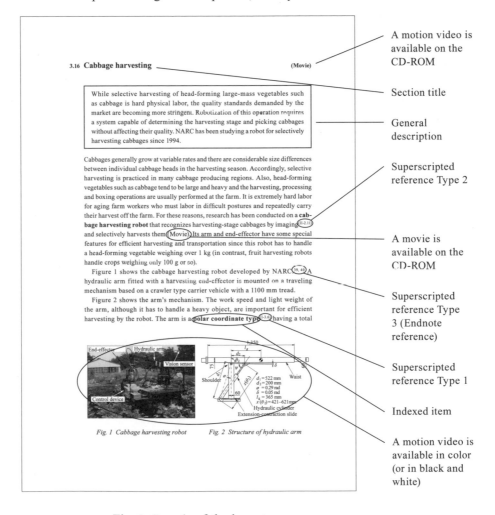

Fig. 1: Sample of the layout

The Japanese original texts of the present project were published in two volumes. The first volume dealt with basic and theoretical issues. The second volume concentrated on the practical mechanisms and applications. This English book is a translation of the second volume.

The superscripted references are made up of three types.

- The first type starts with I followed by two numbers. An example is I-2.4, which means that the readers are invited to refer to section 2.4 of the first volume, which is available in Japanese.
- The second type starts with II followed by two numbers – for instance, II-3.2, which means that the related points are discussed in section 3.2 of this book.
- The third type has neither I nor II at the beginning and points to bibliographic endnotes placed at the end of the volume. The first number is the chapter number and the second the reference item number. For example, 6-9 means that the readers should refer to endnote 9 of chapter 6. The items in bold characters in the text are included in the index at the end of the book.

In order to read the CD-ROM, Acrobat Reader provided by Adobe Systems, Inc. and Quick Time of Apple Computer, Inc. are required. These are available free of charge. Refer to the 'About the CD-ROM' section on the previous page.

Insert the CD-ROM into the computer and click Agrirobo.pdf. The main page (Figure 2) will then start. Once you click on Table of contents (Figure 3) followed by Section (Figure 4), you can then see the list of figures/photos as well as motion videos (without audio) (Figure 5). After clicking Index in Figure 2, you can see the list of indexed items and jump to the relevant figures, photos and motion videos. The initial screen of the CD-ROM shows a full page without displaying tool and menu bars. However, you can change the display style by using Escape, F8 and F9 keys. For details of various functions, refer to the Help of Acrobat Reader.

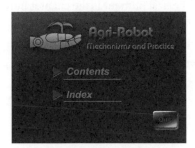

Fig. 2 Main menu

Fig. 3 Table of contents

Fig. 4 Section headings

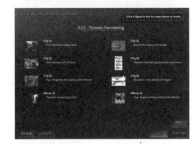

Fig. 5 Specific photos, figures and motion videos

Fig. 6 Enlarged color photos, figures or motion videos

Preface

Agricultural products are in many respects amazingly complex. Take the humble tomato, for example. It ranges in size from 3S to 3L (or more); its color varies from green to red depending on the season or environmental conditions; and in some cases a single fruit presents a variety of colors. Those affected by a pest or disease tend to present even more subtle color variations. Furthermore, many tomatoes that are smaller than medium size have an oblong bullet shape that is so different from the usual image of tomatoes as round fruits that one might suspect that they are a different fruit altogether. There are also many tomatoes, both small and large, with triangular or square shapes that are not found in ordinary supermarkets. Looking at the calyx, there are infinite varieties to its size, color, number of leaves and shape, and each calyx appears completely different immediately, several hours or several days after harvesting.

That agricultural products present vast differences that are unimaginable for ordinary industrial products may seem to be a matter of common knowledge, but in fact, it is only a matter of concern to people who are experts on individual products. Agricultural robots must cater for such diversity and variety. They need to be able to handle delicate fruit such as a peach in a gentle manner. They must be able to operate in the dust of peach fuzz or pear bloom, for example. Yet they also need to be protected from dust and sudden rains in the outdoor environments of agricultural production. They are often exposed to high temperatures and high humidity in greenhouses. Field crops are even more susceptible to variations. The numbers of items and cultivars from field crops increase each year and their physical properties are constantly changing. Agricultural robots are evolving in response to the changing variety and complexity of these agricultural products and field crops. The co-evolution of robots and their objects, as well as encounters with the unpredictable behavior of agricultural products and crops is what makes agricultural robotics so much more challenging, and thus more attractive, than other areas of robotics.

The preceding publication, *Agri-robot (I) – Fundamentals and Theory*, described the basics of the machine vision systems, end-effectors and arms and traveling mechanisms of agricultural robots. Re-reading it one year after publication, the authors became keenly aware of the difficulty in explaining the fundamentals of robot components that are used for a wide range of objects and operations within limited space. Hence, the present book incorporates as many examples of actual research and development as possible and uses many diagrams and photos to explain them for ease of understanding. Moreover, the photos, diagrams and moving images are available on the accompanying CD-ROM so that readers can see them in color and enlarge them for better understanding.

The articles in this book were written by leading researchers in the relevant fields. Chapters 1 and 2 were edited mainly by Kondo, Chapter 3 by Monta, and Chapter 4 by Noguchi. The contributors (in the order of writing) are: Kondo: 1.1, 1.8, 1.10, 1.11, 2.1, 2.3, 2.8, 2.12–2.31, 3.1, 3.8, 3.10, 3.17, 3.20, 3.22; Ogawa: 1.2, 1.5, 2.2, 2.12, 2.14–2.16, 2.24, 2.32, 2.33; Yukumoto: 1.3, 1.7, 1.14, 4.4–4.6, 4.9–4.11, 4.14, 4.16–4.19; Matsuo: 1.4, 1.12, 3.18, 4.1–4.3, 4.7, 4.9, 4.12, 4.13; Arima: 1.6, 1.9, 2.9, 2.10, 3.2, 3.3, 3.9, 3.13, 3.14, 3.16, 3.24; Noguchi: 1.13, 2.5, 2.34, 4.8, 4.15, 4.20–4.24; Fujiura: 2.3, 2.4, 2.7, 2.11, 3.4, 3.12, 3.15, 3.22, 3.23; Monta: 2.6, 3.2, 3.5–3.7, 3.11, 3.19, 3.21, 3.25.

The authors hope that, like its predecessor, this book will be used not only as a textbook or supplementary reader at universities and colleges but also as a reference book to inspire researchers and engineers in research organizations and corporations and to inform those who are involved in the introduction and use of agricultural robots and facilities such as agricultural advisors, corporations and producers of technology.

Many corporations, research centers and universities in Japan and around the world have kindly provided many color photographs, diagrams and moving images for this publication. It would have been impossible to compile so many examples into a single book without these materials. The authors sincerely thank these contributors for their cooperation. We also thank Corona Publishing Co., Ltd. for their assistance in publishing *Agri-Robot (I) – Fundamentals and Theory* and *Agri-Robot (II) – Mechanisms and Practice*.

Further, Dr. Mitsuji Monta, one of our editors, used his talents in preparing and drawing plans, preparing and editing the CD-ROM and doing the layout of this book and dedicated great effort to its publication. His efforts helped us enormously in completing the book in a short period of time and minimizing the cost of editing. Ms Noriko Nishizaki, a graduate student at the Bioproduction Systems Engineering Laboratory, the Graduate School of Environmental Science, Okayama University, assisted us with the laborious process of cutting, pasting and linking diagrams and photos to the CD-ROM. We acknowledge and thank them both for their contributions.

April 2006
Naoshi Kondo

Authors

In the order of writing, as at April 2010:
- Naoshi Kondo (Kyoto University)
- Yuichi Ogawa (Kyoto University)
- Osamu Yukumoto (NARO, National Agriculture and Food Research Organization)
- Yosuke Matsuo (NARO, BRAIN, Bio-oriented Technology Research Advancement Institution)
- Seiichi Arima (Ehime University)
- Noboru Noguchi (Hokkaido University)
- Tateshi Fujiura (Osaka Prefecture University)
- Mitsuji Monta (Okayama University).

Editors

- Naoshi Kondo
  1984, MS from the Graduate School of Agriculture, Kyoto University
  1985, Assistant Professor, Faculty of Agriculture, Okayama University
  1988, Ph.D. from the Graduate School of Agriculture, Kyoto University
  1991, Assistant Professor, The graduate school of Natural Science and Technology, Okayama University
  1993, Associate Prof. Field Science Center, Faculty of Agriculture, Okayama University
  1997, Associate Prof. Faculty of Agriculture, Okayama University
  2000, Manager, Department of Technology Development, Ishii Industry Co., Ltd.
  2003, Director, Ishii Industry Co. Ltd. (2004, Name of Ishii Industry Co. Ltd. was changed into SI Seiko Co., Ltd.)
  2004, Technical Adviser, SI Seiko Co., Ltd.
  2006, Professor, Faculty of Engineering, Ehime University
  2007, Professor, Graduate School of Agriculture, Kyoto University

- Mitsuji Monta
  1988, MS from the Graduate School of Agriculture, Okayama University
  1988, Engineer, Kubota Corporation
  1991, Assistant Professor, Faculty of Agriculture, Okayama University
  1999, Ph.D from the Graduate School of Agriculture, Kyoto University
  2000, Associate Prof. Faculty of Agriculture, Okayama University

2005, Associate Prof. Graduate School of Environmental Science, Okayama University

2006, Professor, Graduate School of Environmental Science, Okayama University

• Noboru Noguchi
   1987, MS from Graduate School of Agriculture, Hokkaido University
   1990, Ph.D. from Graduate School of Agriculture, Hokkaido University
   1990, Assistant Professor, Faculty of Agriculture, Hokkaido University
   1997, Associate Professor, Graduate School of Agriculture, Hokkaido University
   2004, Professor, Graduate School of Agriculture, Hokkaido University

1 Mechanisms of Agricultural Machinery and Emerging Agri-Robots

Advances in computer technology, image processing and mechatronics is heralding the evolution of agricultural machinery toward agricultural robots. Some machines have already been commercialized under the label 'robot'. This chapter describes the current state of progress of practical application and research and the future outlook in the field of agricultural robotics. It also explains the mechanisms of existing agricultural machinery, categorized by work type.

1.1 Emerging agri-robots

While different farm work procedures are used for different farm products, the current farm work procedure can be summarized in very broad terms, as in Figure 1. The current state of development of new agricultural machinery and the outlook for robotization are explained in this section with regard to each type of farm work shown in Figure 1 mainly based on the outcomes of the Urgent Development Project for Agricultural Machinery (1993–1997; 'Urgent Project') and the Urgent Development Project for Agricultural Machinery in the 21st Century (1998–2002; 'Urgent Project 21') conducted by the Bio-Oriented Technology Research Advancement Institution ('BRAIN') of the National Agriculture and Food Research Organization ('NARO') of Japan.

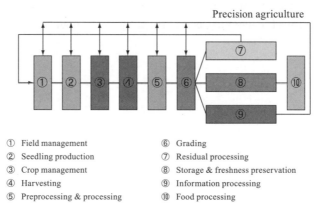

① Field management	⑥ Grading
② Seedling production	⑦ Residual processing
③ Crop management	⑧ Storage & freshness preservation
④ Harvesting	⑨ Information processing
⑤ Preprocessing & processing	⑩ Food processing

Fig. 1 Farm work procedure

In the area of field management, tractor-based **tillage robot**[II-4.9] technology is fairly advanced and almost ready for practical use as reported by the Urgent Project.[1] Since the costs of **GPS**[I-4.1] and some other components have fallen to reasonable levels, it is highly likely that **robot tractors**[II-4.8] will be introduced to farms of certain operational scale once the issue of safe operation is resolved. In fact, many large American tractors are already equipped with a steering assist device[2] and capable of straight-line driving and night-time operation. These devices minimize the waste of fertilizer, agricultural chemicals, seeds and fuel.

In other developments, in 2004, one manufacturer began trials of a **soil sensor** (underground sensor)[2] which was designed to provide field data. This sensor attaches to the **three-point linkage hitch system** of a tractor and is able to provide real-time measurement of moisture, organic matter, nitrogen, **pH** and **EC** (electric conductance) levels.[II-2.1] Over 100 sensors for the measurement of EC only or EC and pH have already been put to practical use in the U.S. and other countries. Sensors capable of measuring organic matter and nitrogen levels are expected to be introduced in Japan as well as large-scale farms in the U.S. and Europe in the near future.

At the same time, soil sampling devices[1] have been developed by the Urgent Development Project for Next-Generation Agricultural Machinery (2003–2007, 'Next-Generation Project'), including devices that collect soil samples for analysis from various parts of a field and record sampling locations by GPS data, automatically crushing and screening soil samples, and performing simple analysis of collected field-moist soil samples for **nitrate nitrogen**, potassium, pH and EC. The Urgent Project 21 has already developed a fertilizing machine[(II-1.3)] that precisely distributes fertilizer based on the results of soil analysis.[1] It is capable of changing the rate of application of basal fertilizer or topdressing gradually while traveling and can be fitted to a **rice transplanter** or a **paddy field cultivator**. The concept of **precision farming**,[4] in which fertilizer and chemicals are applied in quantities as and where needed, is increasingly accepted around the world.

Seedling production work includes seeding, raising seedlings, thinning, grafting, cutting and transplanting. Seeding of vegetables and rice has been automated from the automatic tray feeding stage to soil bedding, firming, covering and watering. A seeder machine[1] has been developed for large seeds such as pumpkin seed which can line up all the seeds in a uniform orientation and germ position using adsorption. Tractor-mounted seeder machines have been in practical use for many years.[(II-1.4)] For seedling raising and grafting work, **cell tray** seedling transplanters[(II-3.1)] equipped with a TV camera[(I-2.3)] or **photoelectric sensors**[(I-3.4)] have been in practical use in other countries for a long time but the Urgent Project undertook the development of a commercial model,[1] which produced a grafting robot for cucurbits.[1] [(II-3.3)] For cutting work, research had been conducted for a fully automated process for chrysanthemum cutting[5–9] before the Urgent Project 21 developed a semi-automatic machine[1] [(II-3.2)] which removed lower leaves and planted ten cuttings after the cuttings were fed into the machine manually. For transplanting work, research and development of tray-to-tray transplanting robots have been underway for a long time and new transplanters for leaf vegetable, strawberry and sweet potato seedlings[1] [(II-1.6)] have been added to the lineup.

In a **plant factory** setting, attempts to robotize mericlone seedling production have been underway. Some machines for seedling propagation have been in development but most of them appear to be still in the research phase. **Seedling planting robots** and **spacing robots** (for crop space adjustment) are also under development.[10(II-3.24)] As plant factories come into practical use in the future, many types of robot will also come into active service.

For crop management work, a machine to cultivate, top dress and pest-control vegetables such as cabbage,[1] a vehicle to fertilize, weed and pest-control rice paddies,[1] and a tractor-mounted machine to weed and cultivate orchards[1] have been developed.

[II-1.7] **Pest control robots**[II-4.13] were commercialized by various companies many years ago in order to avoid human exposure to chemicals. Lately, a remote control type robot equipped with white LED lighting and **CCD cameras**[11] has been developed. The Urgent Project 21 has developed a **multipurpose monorail system** for chemical and fertilizer spraying.[II-4.18] A crop growth measuring device[1] that measures leaf color has also been developed as part of precision crop management.

For harvesting work, the development of various harvester robots[13] has been underway mainly at research organizations, the oldest of which is research on a **tomato harvesting robot**[12] that began over twenty years ago. The harvester robot that is closest to practical application now is one for fruit vegetables grown on **table top culture**[I-3.3] such as strawberries.[II-3.10] However, there are considerable regional variations in the cultivation pattern, plant variety, scale of operation, purpose of cultivation and what is expected of the robot. For this reason, **tailor-made robots** are required for harvesting, customized to local needs and built to regional specifications, as is the case with a **fruit grading facility**, which will be discussed later. Harvesters for lettuce, *daikon* radish, burdock, Chinese cabbage, cabbage, non-head forming vegetables, spring onion and potato[1] [II-1.10] have been developed by the Urgent Project. A device capable of measuring moisture and mass[1] while a **combine harvester** is harvesting grain crops has also been developed. This enables the preparation of an accurate **yield map** of a finely divided field.

Preprocessing and grading work is one of the areas where robotization and mechatronization are most advanced: citrus **fruit grading machines** have been in widespread use at JA and other **cooperative fruit grading facilities**. Such facilities, not only for citrus, are custom-made plants which take account of local strategic considerations. As such, they are subject to complex processes which include: the assessment of the locality, the design of the machine's performance, function and capacity based on local characteristics such as cultivated products, varieties, climatic condition, geographical condition and social conditions. Since they have a large impact on the income of the local community and producers, careful planning involving the whole town is essential in some cases.

Citrus fruit grading facilities have been undergoing mechatronization[II-1.11] for some time with the introduction of image processing technology and near infrared spectroscopy. **Fruit grading robots** with arms[14, 15] for peach, *nashi* pear and apple were introduced in 2002.[II-3.20] For boxing of mostly spherical fruits, **Cartesian coordinate type boxing robots** with end-effectors consisting of **suction cups**[I-3.5] were introduced. Work inside a fruit grading facility involves **depalletizing** and **palletizing** (i.e., off-loading and up-loading containers of produce and shipping boxes on pallets) for which **articulated type general industrial robots** are used.

For vegetables, the Urgent Project 21 has developed[1] a machine which recognizes cabbage and other head forming leafy vegetables by color, adjusts the cutting position and cuts off outer leaves, a device to cut spinach and other soft vegetables at their root, remove lower leaves and bind them, and a device to perform precision root-level cutting and peeling of long spring onion (long white leek). Leek preprocessing and **grading robots**[II-2.26] capable of processing 10,000 leeks per hour were introduced to cooperative grading facilities in 2002.[16] A grading facility capable of full color imaging of elongated fruits such as egg plant and inspecting Japanese quince was also introduced in 2002.[17] [II-2.21]

In animal husbandry, machines such as **mowers**, **tedders**, **rakes** and **balers**[II-1.12] have been in use for a long time. Research[18] on the robot use of GPS and image processing to automatically harvest grass[II-4.15] was conducted simultaneously with research on the aforementioned **automatic guidance tractor** as part of **automatic guidance** research. **Weeding robots** that identify weeds based on reflectance and spot-sprays herbicide have been developed for pasture[1] and algorisms for distinguishing weeds from lawn based on differences in **texture** through image processing is being studied.[19–21] A system to feed optimum amounts of roughage and concentrate to dairy cattle based on milk production characteristics[1] and a **milking robot**[1] have been developed.[II-3.23] This milking robot carries out automated udder washing, drying, **teat cup** handling and milking for a stanchion stall of approximately fifty animals. According to the National Institute of Livestock and Grassland Science of NARO, a total of 130 milking robots have been introduced by various companies in Japan and around 3,000 units have been sold world-wide. As consumers are increasingly concerned about the security and safety of milk products, the development of a high precision sensing system for testing milk for defects and foreign objects is desirable. A **wool shearing robot**[22] was developed in Australia.[II-3.22]

After the work described above is completed, products often undergo further processing before reaching consumers. A diverse range of mechanical systems have been developed for this stage of processing for various products such as brewed foods, juices, pickles, noodles, dairy foods, hams and sausages. They are too diverse and numerous to address systematically in this volume.

As discussed in *Agri-Robot (I) – Fundamentals and Theory*, the roles of agricultural machinery and robots at these stages of work have changed with the times.[I-1.2] While the main purposes of agricultural robots in the twentieth century included labor substitution, product quality enhancement, product standardization and germ-free facilities, accurate provision, recording and utilization of work data has been added to the main roles of the **third generation agricultural robots** in the twenty first century as the sensing function and large memory capacity of the robot means that it can provide information to cater for the new need for food safety and diverse consumer tastes.[23, 24] In particular, the early commercialization of grading robots for various products in various localities is highly anticipated since they are capable of adding and accumulating information which is useful for enhancing commercial value of agricultural produce, hence income, and providing product quality data to be used as farming support information for subsequent years.

Figure 2 shows the type of information that can be obtained from robots at each work stage described in Figure 1, summarized from the perspective of reducing environmental burden and enhancing profitability. It suggests that the soil sensor and the **field management robot** gather field information and are especially important for optimal fertilizer application and that the grading robot gathers information about farm produce and is important for profitability. Management of these two important types of information is essential for environmental load reduction and profitability. It is anticipated that organizations such as JA, which is undergoing a consolidation process, will take on the role of local information center (e.g., soil analysis service center, agricultural product information center etc.) in the near future.

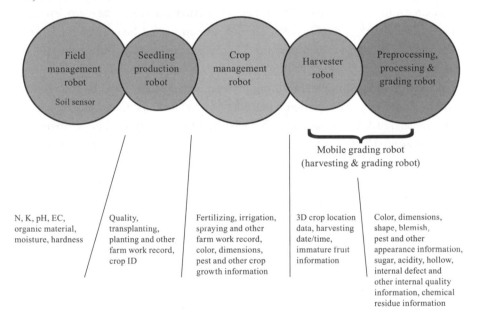

Fig. 2 Information from robots and sensors

According to Shibusawa (2003), 'the Japanese model of **community-based precision agriculture** is a new local farming system which pursues a harmonious mix of environmental protection, profitability in agriculture and food safety as it manages the whole system of production, distribution and consumption by creating information-intensive farms and information-intensive products with the help of sophisticated farming groups and technological platforms.'[25] The soil sensor and the grading robot play an important role in developing farms and products which generate such information (Figure 3).

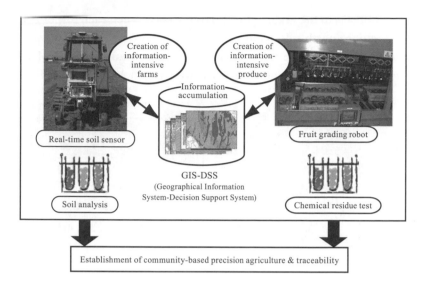

Fig. 3 Community-based precision agriculture

With regard to other robots, it is important that information about agricultural chemical spraying, pest and diseases and crop growth generated by the **crop management robot** is managed properly at the information center. While chemical spraying should ideally be limited to detected areas of infestation and their surrounds, information about the amounts and times of chemical use becomes essential for **traceability** as well as for spraying for general or preventative pest control. In the case of the **seedling production robot**, it would be more practical to manage the information about seedling varieties by using **IC tags** in many cases since seedlings of different varieties may be difficult to differentiate based on their appearance (e.g., different varieties of rice).

From this perspective, a robot arm-type harvester, which has often been considered to be a typical agricultural robot, does not necessarily provide as much information as other robots do and its work rate and work accuracy are, at present, less than, or at best on par with, those of humans in most cases. Consequently, the introduction of harvester robots does not often lead to enhancement of product value or profitability. This is certainly one of the reasons that are deterring the practical application of harvester robots. Hence, it is either the **mobile grading robot**[26(II-2.23)] or the harvesting and grading robot that requires early commercialization in order to achieve precision agriculture. The reason for this is explained next.

Grading robots can attach external appearance information and internal quality information to individual products and packaging based on such information can assure quality, thereby enhancing product value. A mobile form of this grading robot, as shown in Figure 4, can go out to the field with a worker, who passes harvested products to the grading robot. For instance, when each row of fruit trees is divided into the north side and the south side, harvested by hand and graded by the robot as shown in Figure 5, it is possible to prepare **quality maps** (color, shape, dimensions, defect, pest and other maps) as well as a yield map of this field.

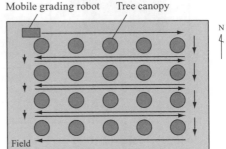

Fig. 4 Mobile grading robot *Fig. 5 Example of mobile fruit grading robot's traveling course*

Like harvester robots, mobile grading robots can gather information about product locations in the field which can be used for tree management. At the same time, it can provide information about the freshness of the produce since it grades harvested fruits on the spot and the time of grading can be regarded more or less as the time of harvesting. In addition, the mobile grading robot can separate graded products from off-grade products on the spot, minimizing the need for storage space because off-grade products can be sent to a processing factory or a biomass plant directly from the field.

Thus, these farm work robots offer many benefits to precision agriculture. Without agricultural robots, it is difficult to manage information accurately or to establish traceability systems in many cases. Current trends in society suggest that the aging of the farming population is inevitable and the number of people returning to agriculture after retirement is increasing. Accordingly, it is essential that robots have human-machine interfaces that are easy and safe for older workers to operate. When harvesting work is also robotized, the harvesting and grading robots will easily figure out which part of the canopy it has picked a particular fruit from based on the positioning of its arm. Further research on these robots is eagerly awaited.

1.2 Tilling machines

> Tilling is an extremely important part of farming work in order to achieve stable yield. The work is usually done by implements such as plow and rotary tiller attached to a tractor. Usual surface tillage involves tilling the top 15–20 cm of field soil.

Tillage means an operation to create an appropriate soil environment for plant growth by optimizing soil conditions such as moisture, hardness, temperature and oxygen levels in the rooting zone by plowing the soil. Specifically, tilling applies a cutting force or a shear force to the soil, and thereby shifts, turns and agitates clods. The operation is generally performed by implements attached to a **tractor** (Figure 1). A tractor is a farm machine that transmits power from the rear **PTO** (power take-off) to drive the implement that it is towing. It is equipped with a **three-point linkage hitch system** to attach the implement and has **hydraulic cylinders**[1-3.3] to lift the implement up and down. Figure 2 shows the PTO and the three-point linkage hitch system at the rear of the tractor. These days, many models are designed so that the implement can be attached and detached while the operator sits in the cab (Movie 1).

Figure 3 depicts a tilling operation. The left photo shows **plowing** to turn soil layers and the right photo shows **rotary tilling**. Usual surface tillage is classified into two types: 'plowing to turn soil layers' by which soil in the ground is plowed and brought up to the surface, broken up and plowed in, and 'rotary tilling' by which surface soil is agitated and broken up. The former uses a set of curved metal plates

Fig. 1 Tractor with rotary tiller (Kubota Co., Ltd.)

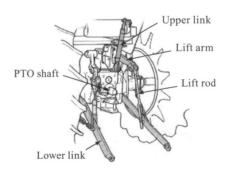

Fig. 2 PTO and three-point linkage hitch at the rear of a tractor

Fig. 3 Plowing for turning of soil layers (L) and rotary tillage (R) (Kubota Co., Ltd.)

or rotating discs called a plow, which is towed by a tractor. The tractor is designed to maintain constant traction in heavy loading operations such as plowing by detecting a load on the linkage and automatically adjusting the height of the implement.

Rotary tilling is performed by **rotary blades** which are rotated by the power transmitted from the PTO of the tractor (Movie 2). The machine has a sensor control system to detect change of engine speed or uneven field surfaces and maintain the engine load and the depth of plowing at constant levels. Some models are equipped with an **inclination sensor** to control the machine for ground inclinations. It detects the inclination of the implement and controls the hydraulic cylinder of the three-point linkage hitch system to keep the implement level. There is also a sensor to detect the angle of the rotary blade cover and control the height of the implement in order to maintain a uniform plowing depth (Figure 4). Rotary blades of various shapes and materials are currently available from manufacturers, which allow farmers to choose appropriate blades for the soil condition of their fields based on durability, turning capacity, and loading. In a plowing method called **down cut**, the blade of a plow cuts into the ground as it is rotated downward whereas the opposite rotation is called **up cutting**. While tilling can be done by either method, they act on the soil differently, and it is said that up cut produces a better result in terms of soil crushing.

There are also **crawler tractors** for wet paddy fields and some models of tractor are capable of adjusting the height of the plow and other implements guided by a laser beam from a laser emitter installed in the field (Movie 3).

● Maintaining a uniform plowing depth in an undulating field

● Microcomputer sensor detects traction loads   ● Microcomputer controls the height of the attachment automatically

Hard soil or rocks   Hard soil or rocks

Fig. 4 Attachment level control

1.3 Fertilizer distributors (manure, powder, granule, and liquid fertilizers) (Movie)

> Various fertilizer distributors are available depending on the type of fertilizer, distribution method and timing. Seeders and transplanters that can apply fertilizer simultaneously are widely used. Variable rate fertilizers and spot applicators for precision agriculture are attracting interest these days.

The mechanism of the **fertilizer distributor** varies depending on the type of fertilizer (manure, compost, granule or liquid fertilizers), distribution method (general application, row application, spot application or in-ground injection), and timing of distribution (basal application or top-dressing)[27] (Movie).

A **manure spreader** (Figure 1) carries very moist and heavy manure and distributes it over the whole field as it loosens coarse fibrous matters. It is usually towed by a tractor and powered by its **PTO**. A manure spreader consists of a tray section, a conveyor to shift manure from the tray to the spreader, and a beater or a distributor rotor to disperse and spread manure evenly.

A **lime sower** (Figure 2) is used for the application of lime, powdered or granuled fertilizer, mainly for basal application, which has a feeder in a relatively low position to minimize the drifting of fine particles and either row-applies or broadcasts the fertilizer. The lime sower has a hopper as wide as the application width, an agitator to prevent bridging and to feed the fertilizer, and a shutter to control the flow. These components are mostly ground-driven in the case of the trailer type or driven by a PTO or a motor in the case of the mounted type.

*Fig. 1 Manure spreader
(Takakita Co., Ltd.)*

*Fig. 2 Lime sower
(Matsuyama Co., Ltd.)*

Spinner type (rotary impeller) or spout type (swinging cylinder) broadcasters (Figure 3) are used for broadcasting granule fertilizers. The broadcaster consists of a hopper, an agitator, a flow control shutter, and a spinner or a spout, and uses the shutter to regulate the flow of fertilizer and feed it onto the spinner or the spout, which provides kinetic energy to release and spread the fertilizer. It distributes fertilizer to

*Fig. 3 Broadcaster
(Matsuyama Co., Ltd.)*

*Fig. 4 Slurry spreader (Star Farm
Machinery Mfg. Co., Ltd.)*

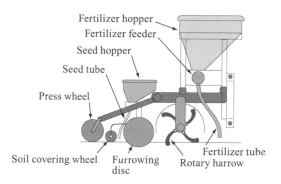

Fig. 5 Seeder with fertilizer applicator (pattern diagram)

a wide area but is easily affected by the wind. Many are tractor-mounted, some are towed by a tractor or a walking tractor, and some are self-propelled.

A row applicator is used for row application of granule fertilizer and has a hopper, a shutter, a roll type or perforated plate type feeder, and an applicator tube.

For liquid fertilizers such as slurry, a **slurry spreader** (Figure 4) is used for broadcasting the whole field and a slurry injector is used for direct in-ground application. The slurry spreader has a traveling section, a tank, a pump (a vacuum pump in many cases), and an applicator section. The applicator section can be a jet nozzle type, impact plate type, impeller type or spinner type. The slurry injector has chisels and nozzles in the applicator section which rip up the soil and inject the slurry fertilizer.

Machines to apply basal fertilizer simultaneously with other works such as transplanting or sowing are also widely used. A **row side applicator** is attached to a **rice transplanter** and places granular or paste fertilizer on the soil next to seedlings that are planted at the same time. A **fertilizer drill** (Figure 5) is comprised of a harrowing and furrowing section, a row fertilizer applicator, a seeder and a press

Variable rate fertilizer display/consol panel

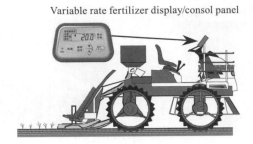

Fig. 6 Variable rate fertilizer (basal fertilizer attached to transplanter) (BRAIN)

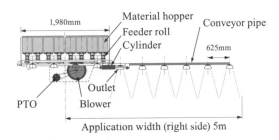

Fig. 7 Variable rate fertilizer (NARC, Hokuriku Research Center)

wheel for each row and is capable of harrowing, furrowing, fertilizing, seeding, soil covering and firming in a single operation.

Active research has been progressing on precision **variable rate fertilizers** (Figures 6 & 7) for fertilizer management in **precision agriculture** which are expected to reduce excessive fertilizer use and environmental burden.[28, 29] There are two ways to determine the rate of fertilizer application: one is based on a fertilizing map prepared from field and crop information (map-based) and the other is based on crop information obtained by a sensor at the time of fertilizer application (sensor-based). The map-based system uses geographical information from **GPS**[1-4.1] in mesh form. The precision variable rate fertilizer controls the feeder mechanism with a servo motor based on signals from a computer or manually operated by the operator in many cases.

Seeders (broadcast, drill, hill-drop)

> Seeders are machines used to sow various crop seeds in nursery beds, seedling boxes or directly in fields. Since a majority of seeding mechanisms are automated, automatic guidance of the tractor or the base vehicle as its driving force is the key to its robotization.

The action (function) of the **seeder** usually entails leveling, furrow opening and dibbling to prepare the condition of the seeding site, followed by feeding and sowing of designated amount of seed, soil covering and firming. Many seeders perform tilling and seeding simultaneously as well as ridge making, mulching and chemical spraying.

Seeders for seedling boxes and **cell trays** are mostly a stationary type used inside a nursery plant which performs seeding work on its own. Seeders used in fields, however, cannot perform the work on their own, but must obtain power for traveling and operating from a person pushing the machine, a walking type tractor or a riding type tractor (collectively called the 'tractor') to perform seeding work. The types and characteristics of seeders by crop or seed form are explained below.[30, 31]

(1) **Vegetable seeder**: performs dibble sowing of raw seeds of vegetables such as *daikon* radish, burdock and spinach or coated vegetable seed, inoculating several seed grains at a time at equal intervals, or drill sowing. The seed feeder mechanisms include the roller, belt, perforated plate and vacuum types. Figure 1 shows a vacuum type seeder with a vacuum pump unit (center of figure, backpack type), which creates negative pressure to draw a designated number of seeds (one to several) to the end of radially set nozzles in the seeder unit (bottom right of figure) and inoculates them at equal intervals.

(2) **Seeder mulcher**: This machine inoculates several grains of *daikon* radish and carrot seed in each seed hole as it performs ridge making and hole mulch laying. A tractor driven type also performs tilling and some models can be fitted with a sprayer which applies agricultural chemicals over seed holes at the same time.

(3) **Tape seeder**: Mainly for seeding vegetables in the field or nursery beds, the seeder machine makes seed tape by sealing several seed grains in water soluble film or non-woven fabric tape. The tape seeder is a machine to bury the seed tape in the soil.

(4) **Drill seeder** and **rotary seeder**: This machine sows wheat, soy, vegetable and grass seeds in stripes or drills and many of them are equipped with a fertilizing or chemical spraying device. Many of the grain drills for soy and grass seeds are multiple types with more than ten drill rows, attached to a tractor which also tills and fertilizes. Figure 2 shows the rotary seeder mechanism.

Fig. 1 Vacuum seeding machine for vegetables (hand seeder, single row drill)
(Keibunsha Seisakusho)

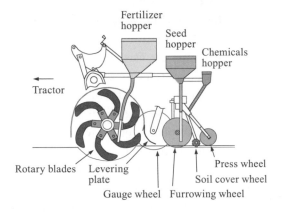

Fig. 2 Mechanism of rotary seeder

Fig. 3 Precision drill seeder for direct
sowing of rice (BRAIN)

Fig. 4 Aligning seeder for large
and elliptical seeds
(Yanmar Co., Ltd., BRAIN)

(5) **Direct seeder** for rice: This machine sows seed rice directly in dry or wet paddies instead of transplanting rice seedlings. There are both drill and broadcaster types. A combination of a tractor and a rotary seeder is generally used for direct dry paddy sowing. A special seeder unit for **coated seed** attached to the power/driving unit of a **riding type rice transplanter** or a **paddy management vehicle** is used for direct seeding on wet paddies.

(6) **Potato planter**: sows seed potatoes. Its basic functions include ridge making, seeding, soil covering and pressing. Some models are equipped with devices to cut seed potatoes to a uniform size or to apply fertilizer. They use various feeding and seeding mechanisms but many require some manual operation such as feeding the seed potatoes into the machine.

(7) **Nursery seeder**: Most seeders that inoculate seedling boxes and cell trays inside nurseries are stationary types. System development and automation, including the pre-seeding and post-seeding stages, is at an advanced stage. Some machines for rice nursery seeding can perform seedling box feeding, soil packing, irrigation, sowing, soil covering and seedling box shifting (to the germinator). Seeders for vegetable nurseries are required to be able to inoculate a single seed of various vegetables at the center of each seedling pot or each cell of a cell tray to a uniform depth. They adopt various mechanisms and capabilities (Movie). Some seeders have been developed that are capable of aligning seeds in a particular orientation so that all seedlings have the same germination position and leaf orientation in view of feeding seedlings to grafting machines (Figure 4).

Rice production is one of the crops in which cultivation mechanization is most advanced. Rice transplanters have achieved considerable labor saving in transplanting work, which was previously particularly hard labor. The mechanical planting action which characterizes every rice transplanter is generated by an ingenious combination of gears that is not found in any other agricultural machine. They have also incorporated various measures to maintain the optimal posture of transplanted seedlings.

According to Japan Agricultural Mechanization Association,[32] the total number of shipments of **rice transplanters** was 50,113 in 2002, 48,054 of which were sold in Japan while 2,059 were exported. Of the domestic Japanese shipments, 43,285 were riding type and 4,769 were walking type. Thus, over eighty percent of the 2002 shipments were **riding type rice transplanters.** Manufacturers have produced various models for transplanting between four and ten rows at a time. Prior to 1975, a majority of rice transplanters in Japan were walking types. They used a transplanting mechanism called the crank type, which achieved a beautiful seedling posture by imitating the transplanting action of a human hand. From 1975, the use of riding type rice transplanters spread rapidly throughout the country, leading to further labor saving.[33] One of the technological innovations here was the **rotary type planting mechanism**, which replaced the **crank type planting mechanism** (Movie 1).

The crank type mechanism had a simple construction by which drive shaft revolutions were transmitted through a crankshaft and a four-link mechanism to seedling planting fingers, which performed accurate planting. However, increasing vibrations at higher speeds posed a serious problem. In contrast, the rotary type mechanism cleverly uses **planetary gears** with two fingers attached to one rotating casing. Since the mechanism can plant a seedling twice when the drive shaft revolves once, it is possible to maintain low revolutions while the transplanter travels at higher speeds. As a result, it can plant seedlings with low vibrations even at a high speed (Movie 2).

Figure 1 shows the locus of a planting finger of the rotary type rice transplanter. It is based on a mechanism utilizing planetary gears as shown in Figure 2. Gears A, B and C are called **sun gear**, **idle gear** and **planetary gear** respectively. Their axles are supported by an arm and the three gears are all engaged. In the current example, the planter finger is attached to gear C and gears C and A have the same number of teeth. Where gear A is fixed and the arm is rotated around gear A, gear C orbits around gear A without rotating. Consequently, the planting finger moves

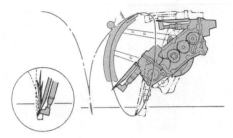

Fig. 1 Planting finger locus of rotary type rice transplanter (Kubota Co., Ltd.)

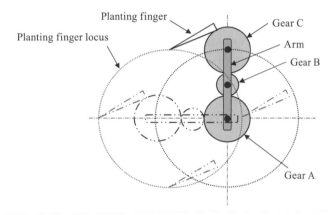

Fig. 2 Mechanism of rice planting finger

in a circle with a radius of the length of the arm while maintaining the same orientation. In reality, eccentric gears or non-circular gears of carefully calculated complex shapes may be used so that the planting finger approaches the seedling tray at a constant angle in order to steadily pick-up a uniform number of seedlings from a mat type nursery. The planting finger is designed to follow a path which makes seedlings stable in the ground at the moment they are inserted in the soil by the planting finger and released. At the same time, the tip of the finger has an auxiliary mechanism to give seedlings a slight push into the ground to facilitate the seedling release and correct the seedling posture. Manufacturers use various forms of push rods based on their unique ideas and operate them at the optimum push speed. The uniform planting locus does not guarantee a uniform seedling planting depth. This is why rice transplanters are equipped with floats which level the paddy surface and sense the depth in order to raise or lower the whole transplanting unit and precisely control the path of the planting finger so that it maintains a uniform distance from the paddy surface.

Fig. 3 Rice transplanter in action
(ISEKI & Co., Ltd.)

Fig. 4 Rice transplanter with fertilizer
applicator (Minoru Industrial
Co., Ltd.)

As rice transplanters become faster and more powerful, the development of power transmission systems has also progressed in recent years. Various systems are being developed, from mechanical transmissions with smart combinations of gears and clutches to, more recently, the Continuous Variable Transmission (**CVT**) capable of gearless drive, the hydraulic type Hydro Static Transmission (**HST**),[34] and the Hydro Mechanical Transmission (**HMT**), which combines mechanical transmission and HST. They aim for labor saving in the area of machine operation for aging farmers and their easy starting and gearless driving mechanisms achieve simplicity of operation. An increasing number of recently developed rice transplanters are capable of other tasks such as fertilizer and chemical application. This is called 'simultaneous fertilizing' or 'simultaneous spraying' which entails application of fertilizer or chemicals near seedlings as they are transplanted and performed by a **fertilizer applicator** or a chemical sprayer mounted on the rice transplanter (Figure 4) (Movie 3). These implements are attachments to the rice transplanter and their operation is synchronized with the transplanting unit. For these fertilizer applicators, which are in particularly widespread use, some manufacturers use the side furrow fertilizer roll type feeder mechanism and others use the perforated plate feeder mechanism.[II-1.3]

Commercial models of vegetable transplanters are available in fully automatic, semi-automatic with manual seedling feeding, riding and walking types. Agricultural machinery manufacturers are actively pursuing joint development and OEM projects in their efforts to adapt their machines to an ever wider range of vegetables.

'Naehansaku' is an old saying frequently uttered by Japanese vegetable farmers. It means that the quality of seedlings determines half of the quality of the harvest, emphasizing the importance of appropriate nursery management for growing high quality seedlings and appropriate transplanting to produce good crops. The posture of transplanted seedlings and the transplanting method are key factors in vegetable cultivation and **vegetable transplanters** can contribute to the uniformity of seedlings and their growth as well as saving labor.

Vegetable transplanters can be broadly divided into two types: the fully automatic type, which picks seedlings out of a nursery tray and plants them in the field, and the semi-automatic type, which automatically plants seedlings that have been manually picked from a nursery tray and fed into the machine. Fully-automatic models can plant seedlings once a nursery tray is set in the transplanter but they can only handle varieties that are raised in **cell trays** or **pulp mold pots**. Semi-automatic models require an operator to feed seedlings and are therefore unsuitable for long hours of continuous planting, but because the operator can check the quality of seedlings at the time of feeding, these machines can plant a relatively wide variety of seedlings.

Figure 1 shows a fully **automatic transplanter** (**plug seedling** transplanter) (Movie 1). This transplanter consists of a seedling feeder unit, a transplanting unit and a traveling unit. The feeder unit consists of a traverse feeder device which shifts a tray on the seedling table laterally by one plant at a time and a longitudinal feeder device which shifts a tray vertically by one row at a time. It can handle cell trays with either 128 cavities or 200 cavities by moving a lever. The transplanting unit thrusts its seedling picking fingers, shown in Figure 2, through the leaves of a seedling, inserts them into its cell, grabs it at the soil molding and transfers it to the planting unit. The seedling is held in a hopper type planting device, which descends and makes a seedling hole at the top of a ridge at the same time. The bottom of the hopper opens and rises, leaving the seedling in the hole. A pressing roller behind the hopper gathers soil at the base of the seedling to cover it. The hopper type planting device uses an intermittent motion mechanism to ensure that the optimum seedling posture is maintained when the traveling speed or the plant spacing varies.

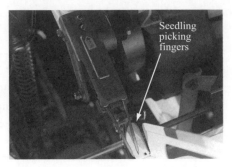

Fig. 1 Automatic vegetable transplanter (cell mold seedlings) (ISEKI & Co., Ltd.)

Fig. 2 Seedling picking fingers of the seedling feeder unit

Semi-automatic transplanters (Movie 2) usually have a turntable with cups arranged in a circle. Seedlings are fed into the cups by hand, and then planted using a device similar to that of the fully automatic type. They can be adapted to both leaf vegetables and fruit vegetables of various forms and sizes by changing the cups and the hopper (Movie 3). Figure 3 shows the first commercial model of a semi-automatic **sweet potato planter** (ISEKI & Co., Ltd.), which employs the concave planting method (Movie 4). Sweet potato seedlings are planted using either the concave planting method or the angular planting method. The former is a complex procedure requiring special tools to plant a long seedling horizontally in a shallow hole but it minimizes interference between potatoes to produce many well-shaped crops. The angular method plants short seedlings closely at an angle to the ground and is used to produce a limited number of large potatoes in a shorter period of time, mainly for processing. Hence, there was a strong demand among potato farmers for transplanters capable of concave planting.

The work procedure is as follows. Seedlings are held down on a continuous conveyor belt by a sponge and a brush. Thus seedlings of any length can be fed into the machine. The seedlings maintain the same posture as they are carried into the planting device on the conveyor. The pincer grip type planting fingers shown in Figure 4 pinch a seedling near its lower end, insert it into the soil and release it at the lowest point. The open fingers ascend and exit the planting hole to return to the initial position. One of the features of this transplanter is that conventional seedlings can be used as they are.

Seedling transplanters for various other vegetables, including cabbage, lettuce, white leak (Movie 5) and onion (Movie 6) have also been commercialized and are playing a part in the development of local industries throughout Japan.

*Fig. 3 Semi-automatic sweet potato
transplanter (ISEKI & Co., Ltd.)*

*Fig. 4 Pincer-grip type
planting device*

> For cultivating, cultivators and rotary cultivators are used in upland fields and rotary cultivator weeders are used in paddy fields. Precision cultivator weeders for paddy fields have been developed recently. Brush cutters are used along ridges or causeways.

In upland fields, the soil in interrow spaces (between ridges) is fluffed up and softened by shallow plowing for the purposes of improved aeration, moisture regulation and improved fertilizer breakdown as well as mechanical weeding during the growth period. This work is called cultivation. **Cultivators**, **rotary cultivators** and **weeders** are used as **cultivating machines**. Some models are equipped with a **fertilizer applicator** or a ridger to fertilize and make ridges simultaneously.[35]

A cultivator (Figure 1) consists of various components including a frame, weeding knives (fingers), fittings and ruler wheels. It works three to five rows at a time with a few cultivating/weeding knives for each row. Many are directly mounted on a tractor while some are towed. It carries out interrow cultivation and weeding as it breaks the ground surface but some weeds are left behind, inside rows, and have to be removed by hand. A rotary cultivator (Figure 2) has a frame with two to five rotary units which is attached to a tractor's **three-point linkage hitch system** and powered by its **PTO**. It uses the turning and agitating motions of its rotary blades to carry out interrow weeding by tilling between rows, pulling out, cutting and burying weeds under scattered soil. A weeder has a frame with many **spring teeth**, which pull out weeds as they enter and exit the soil by their spring action.

A walker type **paddy field cultivator** (Figure 3) is used for cultivation of paddy fields. It uses a small engine to drive rotors with weeding rotors, which cultivate,

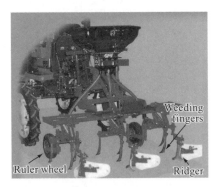

Fig. 1 Cultivator with fertilizer applicator and ridger (Kubota Co., Ltd.)

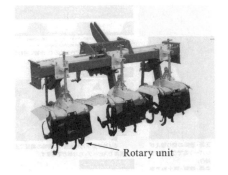

Fig. 2 Rotary cultivator

Fig. 3 Cultivator for paddy field
(Kioritz Corporation)

Fig. 4 Precision weeder for paddy field
(BRAIN, Kubota Co., Ltd.)

agitate and turn the interrow soil as it glides forward. It has grass dividers in front of the weeding rotors in order to avoid injuring rice plants and straight-movement sleds behind them in order to ensure running stability and to control cultivating depth. It generally covers two to four rows. More recently, a **precision** weeder **for paddy fields** (Figure 4) has been developed and commercialized. It uses rotors for interrow weeding and oscillating **rakes** for intrarow weeding to carry out simultaneous weeding of interrow and intrarow spaces with high efficiency that had previously been impossible.[36, 37] The operating efficiency of an eight-row model is reportedly five times that of a three-row walker type cultivator for paddy fields.

A **cultivator for orchards** (Figure 5) (Movie), developed by BRAIN, is attached to the right side of the tractor through a three-point linkage hitch system. Since the implement is laterally extendable and retractable, it can work around tree trunks. The implement is easily detachable and other attachments are available

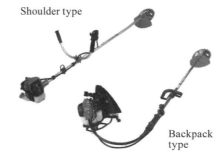

Shoulder type

Backpack
type

Fig. 5 Cultivator for orchard (BRAIN)

Fig. 6 Brush cutter (Kioritz
Corporation)

for grass cutting and cultivation. When an additional **offset mower around trunk** is attached, the machine can also cut grass around tree trunks, as well as earthing up and ridging.

Compact brush cutters (Figure 6) are used for weeding paddy field ridges or causeways. They are either shoulder type or backpack type and consist of an engine, a centrifugal clutch, a control lever, a gear box and a cutting attachment. With the backpack type, operators carry the engine on their back and power is transmitted to the cutter through a flexible shaft. They are usually powered by an air-cooled two-cycle engine with a displacement of 20–40 ml and the cutter is a **circular saw** type or **plastic cord cutter** type.

Thinning is done to ensure an appropriate planting density by removing some seedlings where seeds have been sowed more densely than necessary in order to achieve a required establishment rate. A thinning machine cuts or pulls out excess seedlings along crop rows or hills using rotating or oscillating thinning fingers. The thinning fingers are driven by the PTO shaft or ground wheels. When a cultivator is used for thinning work, weeding blades of appropriate width are attached to its tool bar and the machine is driven at a right angle to crop rows. In recent years, research has been conducted on selective thinning systems which use image processing to identify and evaluate vegetable seedlings.[38]

Various chemical sprayers have been developed for grain, vegetable and fruit crops, including sprayers, mist blowers, fog machines, dusters and granule applicators for a range of purposes. Aerial application using helicopters has recently become common. The development of machines capable of keeping accurate records of chemical usage is awaited.

In the process of crop cultivation, the growth of plants is often hindered by pests and pathogenic microbes. Various control methods have been developed to combat this problem. While the use of light traps and attractants and the utilization of insects and animals as natural predators, for example, have been put to practical use, the most effective and widely used method, regardless of place and time, is chemical control. However, this method has problems such as pesticide residue and harmful effects on workers and, combined with the growing popularity of organic farming, the minimization of chemical use and the full-automation of spraying procedures is highly desirable. The traceability of information about chemical spraying is of the utmost importance and the development of machines that are capable of accurately recording the time and place of spraying and the name and amount of chemicals sprayed is urgently required.

Sprayers using a pump to draw liquid chemicals into a tank and squirt atomized liquid globules out of a nozzle have been used for a long time. Various nozzles with different internal structures have been developed for different particle diameter, spray travel and spray coverage requirements. A machine with a boom with many nozzle attachments and mounted on a tractor together with chemical tanks and **power sprayers** to achieve wide spray coverage is called a **boom sprayer**.[II-4.8] The boom can be retracted and vertically stored when the vehicle travels outside of fields. A **mist blower** atomizes liquid chemicals into smaller particles and uses a fan so that the mist covers a greater distance. A self-propelled vehicle mounted with a multiple nozzle mist blower, large fans, tanks and pressure pumps, as shown in Figure 1, is called a **speed sprayer** (Movie 1). A **fog machine** atomizes a fluid into a fog-like vapor which it disseminates throughout a facility. There are self-propelled and stationary type fog machines as shown in Figure 2, most of which operate automatically using a timer switch. Figure 3 shows an electrostatic sprayer which emits a negatively charged fluid in order to improve its coating property (Movie 2). The table[39] below is a list of particle diameters for these **sprayers** using fluid, dust and granular formulations.

A comparison between liquid sprayers and dust sprayers shows that, broadly speaking, liquid particles are easier to control, can be used for multiple purposes

Fig. 1 Speed sprayer (Kioritz Corporation)

Fig. 2 Stationary and self-propelled fog machines (Kioritz Corporation)

Fig. 3 Electrostatic sprayer (Minoru Industrial Co. Ltd.)

Table: Types of sprayer and particle diameters

Sprayer	Chemical	Diameter range (μm)	Average diameter (μm)
Sprayer	Liquid	150–440	200
Mist blower	Liquid	30–100	40
Low/ultra low volume sprayer	Liquid	40–140	70
Fog machine	Liquid	0.5–50	4
Sprinkler	Liquid	500–3,000	2,000
Duster	Dust	0.5–100	20
Granule applicator	Granule	250–1,500	850

and offer better coating properties while dust sprayers are mechanically simpler and weigh less. A **duster** emits powdery chemicals into the air with the help of a blower. Some dusters can be operated by a single worker while boom type blow heads 30–50 m in length are often used for more efficient operation. Granule particles have larger diameters than dust particles and the machine to spray them is called a **granule applicator**. Many dusters can be used as granule applicators or mist blowers by simply replacing nozzle pipes or blower heads.

Aerial spraying of a liquid or dust formulation from radio-controlled **helicopters** (Figure 4) is increasingly common. Paddy field spraying jobs are often subcontracted to JA and other bodies. There are tractor-drawn **soil disinfectors** (Figure 5) and tiller-drawn soil disinfectors.

Non-chemical pest control machines include **insect collectors**, which use a blower to blast insects off crops and into a collector (Figure 6). They are useful in reduced chemical and chemical free farming practices.

Fig. 4 Radio-controlled helicopter with sprayer[40] (Yanmar Agricultural Equipment Co., Ltd.)

Fig. 5 Soil disinfector (Kioritz Corporation)

Fig. 6 Insect collector (Minoru Industrial Co., Ltd.)

There are two basic types of grain harvesting machines: conventional combine harvesters and head-feeding combine harvesters. The latter are also called the Japanese type. The first Japanese made two-row head-feeding combine harvester, which became the prototype for current models, was launched by ISEKI & Co., Ltd. in July 1967. This type of combine harvester is now manufactured and marketed by various farm machinery manufacturers and is widely used.

A **combine harvester** is a harvesting machine which reaps crops such as rice and wheat, threshes grains from heads, separates admixtures such as dust and leaves and collects clean grain kernels while traveling over fields. It consists of a traveling section, a **reaping section** and a **threshing section**. Broadly speaking, there are two types of combine harvesters: **conventional type combine harvesters** (Movie 1) which gather in the whole plant for threshing and **head-feeding type combine harvesters** which thresh heads only.

The conventional combine harvester generally passes reaped crops through a toothed rasp bar cylinder and a concave to thresh grains by impact and friction. This mechanism is widely used in large-scale farms all over the world due to its high processing capacity but the amount of grain loss tends to be high. BRAIN, Yanmar Co., Ltd. and Kubota Corporation jointly developed and commercialized large multipurpose combine harvesters, shown in Figure 1, which are now used by large farms in Japan. They adopt a screw type threshing mechanism which threshes grains by passing harvested culms through a space between a long rotor and a cylindrical concave in spirals in the axial direction. This mechanism tends to require more power but generates less grain loss.

Fig. 1 Multipurpose combine harvester (BRAIN)

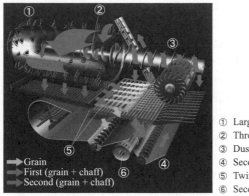

① Large-diameter threshing drum
② Threshing chamber re-threshing drum
③ Dust exhauster re-threshing drum
④ Second auger
⑤ Twin rack
⑥ Second fan

➡ Grain
➡ First (grain + chaff)
➡ Second (grain + chaff)

Fig. 2 Grain threshing and grading section (ISEKI & Co., Ltd.)

The head-feeding type combine harvester cross-feeds uniformly arranged heads and only about 400 mm from the ear tip enters the threshing chamber. This is considered to be a more efficient mechanism than the conventional one with less grain loss and grain damage. Combine manufacturers vie with one another to enhance throughputs of the threshing and grading sections and to increase the operating speed. Figure 2 shows a section of a threshing mechanism. This example has a large-diameter **threshing drum** and two **re-threshing drums** to achieve higher threshing capacity and grading precision. Admixtures such as husks and chaff are threshed in the threshing drum and separated and reprocessed by the first dust exhauster re-threshing drum. The remaining admixtures are recovered by the secondary auger and carried to the second re-threshing drum and returned to the sieves. The grading performance is enhanced by opening and closing the sieves or adjusting the winnowing wind according to the operating condition (Movie 2). Threshed and graded grains are temporarily stored in the grain tank and transferred to truck-mounted grain containers by augers when the tank is full (Movie 3).

Since combine harvesters are mainly used in paddy fields in Japan, the traveling section is usually fitted with rubber **crawlers**. They need to use a **continuously variable transmission** since they must be able to finely adjust the ground speed according to the rice crop condition. Hydrostatic transmissions (**HST**) and belt type continuous variable transmissions are currently used. Many models use a clutch and brake type system by which the operator cuts power to one side and applies the brake to pivot the machine, but this makes a machine unstable. To improve the turning characteristics of the crawlers, a two-pump two-motor system with separate HSTs installed for the left and right crawlers has been adopted. Also, a transmission system has been developed to enable differential turning of the left and right crawlers according to the tilt angle of the steering lever and, as a result, repetitive tight turning maneuvers that are required in small fields have been made easier even in poorly drained paddy fields.

As the name implies, a combine harvester is a combination of the reaping, threshing, traveling and **discharged straw processing units**. Accordingly, many controlling procedures are involved in its operation. Various automatic control devices have been adopted for the purposes of saving labor, operational safety and operability. Direction controllers use a switch inside the grass divider to detect crop rows and automatically steer the machine. Cutting height controllers use **ultrasonic sensors**[1-3.4] to detect the ground height of the harvesting area and keep the reaping unit at an appropriate height. There are controllers to notify the operator when the machine is clogged with straw or the grain tank is full, or control the ground speed according to the processing volumes at the threshing and grading section. Since the attitude of the body affects the machine's reaping and threshing performance, some models are fitted with a controller capable of **horizontal body control** regardless of ground surface inclination. In Western countries, combines are used mainly for corn (Movie 4), soy bean and wheat and equipped with removable headers for different crops.

A **binder** (Figure 3) is a walking type harvester used in terrace paddy fields that are too small for a combine. It was widely used prior to the popularization of combine harvesters. This harvester performs reaping and binding at the same time and consists mainly of a crop lifter, a reaping section, a binding section and a traveling section (Movie 5). The work capacity of a single-row binder is about 7 ares/hour (700 m²/hr) and a two-row binder around 15 ares/hour (1500 m²/hr). Its **binding mechanism**[II-3.21] is used as an attachment for combine harvesters as well.

Fig. 3 Binder (YANMAR Co., Ltd.)

> Various machines have been developed for harvesting varieties of root vegetables, which is hard work in the field, and some leaf vegetables. Harvesters for processing fruits and fruit vegetables have also been developed, and shakers are used in the United States.

(1) **Root vegetable harvesters**: Harvesting root vegetables is hard work since the worker must endure many hours bending down to collect crops growing in the ground and carry out incidental work such as soil and foliage removal. The development and automatization of harvesters is therefore urgently required.

Figure 1 shows an example of a root vegetable harvester. This machine harvests crops such as sweet potato, potato and carrot, carries one to four people, and mainly performs digging, picking, grading, storing and transporting works. Using front-end plows, it digs up underground crops and feeds them onto an elevator. While the elevator carries crops upward, small debris falls down. Crops are carried onto a conveyor belt where they are separated from larger stones and admixtures and graded. Graded crops are put in containers and transported away. The harvester has a self-propelling mechanism and minimizes ground contact pressure by using **crawlers**. The work rate of this harvester is reported to be 2.5–5.5 hours per 10 ares (1000 m^2).

Sugar beet harvesters, **onion harvesters** and *daikon* **radish harvesters** (Movie 1) have also been developed.[41] As yet, though, nothing similar has been developed for crops that are over 50 cm long such as burdock, long carrot and Chinese yam. At this stage, they are dug up with trenchers or root vegetable **plows** and harvested manually.

(2) **Leaf vegetable harvesters**: Mechanized harvesters are eagerly awaited for leaf vegetables such as cabbage, Chinese cabbage, lettuce, spinach and *mitsuba* (Japanese honewort) since this process also requires workers to bend down. Head-forming vegetables such as cabbage, Chinese cabbage and lettuce are generally harvested in several batches as they grow to a certain size. Harvesters for these crops are currently under development.

Figure 2 shows a tractor-mounted **Chinese cabbage harvester**[42] harvesting a field. It consists of a set of biaxial screw augers, pincer-grip belts, rotating disc cutters, an elevator and a rack as shown in Figure 3. The biaxial screw augers hold the base of Chinese cabbages between the left and right grooves, pull them out of the ground and adjust the cutting height. At the same time, outer leaves and bases are cut and transferred to the end of the screw augers to be returned to the field. Pincer-grip belts move with the screw augers to pull out Chinese

Fig. 1 Potato harvester (Kobashi Kogyo Co., Ltd.)

Fig. 2 Chinese cabbage harvester[42]

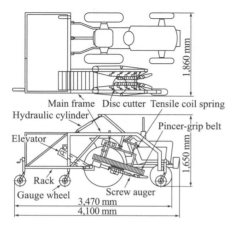

Fig. 3 Mechanism of Chinese cabbage harvester[42]

Fig. 4 Leek harvester (Kobashi Kogyo Co., Ltd.

cabbages and transport them to an elevator after discarding outer leaves and bases. Chinese cabbages are harvested in an upright position, and then they are pushed over sideways by a resister board installed above the end of the grip belts so that the bottom side of the head faces the direction of transport as it is a more stable position when being transported on the elevator. Side-lying Chinese cabbages are transported by the elevator to the rack. The harvesting speed of this machine is around 2.5 seconds per cabbage and the work rate from head cutting to loading onto a truck off field with two operators is 4.9 hours per 10 ares (1000 m²). **Cabbage harvesters** with similar mechanisms have also been developed. Figure 4 shows an **onion harvester**.

(3) **Fruit harvesters**: Harvesters for trench-planted tomatoes for processing were developed many years ago and have been in use in the United States (Figure 5). Harvesters for cherry, apricot, prune, olive, orange and lemon fruits are known

Fig. 5 Harvester of tomato for
processing

Fig. 6 Cherry shaker

as **shakers** (Figure 6) which vibrate fruit trees and collects fruit that drops on to a sheet (Movie 2). Finger type harvesters[41] are used for blueberry and grape. All of these harvesters are for processing fruit.

> A wide range of grading and preprocessing machines for grains, vegetables and fruits have been put into practical use. The introduction of mechatronic grading machines for citrus fruits began in 1997, mainly at JA enterprises in western Japan. The current machines use color TV cameras which take images of fruits from six different directions to grade them.

After harvesting, grains, especially rice, go through stages of drying, husking and milling. Dried grain is stored in a facility called a **country elevator** and hulled rice (brown rice) is stored in a facility called a **rice center**. Please refer to the literature[43] for various types of machinery for each processing stage, including **dryers**, **rice huskers**, **paddy separators** and **rice milling machines**, which have all been around for a long time.

For fruits and vegetables, the **plate type fruit sorting machine** for citrus fruits and the cucumber **shape sorting machine** using black-and-white image processing are classic types. In 2002, the **fruit grading robot**[II-3.20] for deciduous fruits fitted with color **TV cameras**[I-2.3] and nondestructive sensing devices such as **near infrared spectroscopy**-based light sensors came into practical use.[14, 15] **Cartesian coordinate type**[I-3.6] robots which use **suction cups**[I-3.5] to pack graded fruit in boxes (Movie 1) have been developed for various fruits (peach, apple, *nashi* pear, persimmon, tomato, kiwi fruit, cucumber and egg plant) and **articulated type**[I-3.6] robots to palletize boxes of graded fruit (**palletizing robot**) (Movie 2) are already in use. However, since a **fruit grading facility** often means a facility where citrus and other fruits are graded on the conveyor belt, the type of grading facility used for mandarin orange and other citrus fruits, potato, persimmon, tomato and kiwi fruits[44] is explained below.

When palletized containers arrive at the grading facility, **depalletizers**, shown in Figure 1, unload each container from the pallet and carry it to the **fruit grading machine**. **Dumpers**, shown in Figure 2, slowly turn containers upside down and feed fruits onto conveyor belts. Fruits are cleaned with brushes or water or prepared with wax as required and carried to the inspection section shown in Figure 3. It is important here that a large amount of fruits are brought into a single line and transferred to another conveyor belt running at a different speed so that fruits are ultimately spaced at intervals of 15–20 cm (Figure 4). This spacing ensures smooth flow of image processing and internal quality inspection. If some fruits are not spaced appropriately when they pass through the inspection section, they are detected by the imaging device, diverted to another line and returned to the lining-up stage.

Fig. 1 *Depalletizer (SI Seiko Co., Ltd.)*

Fig. 2 *Dumper (SI Seiko Co., Ltd.)*

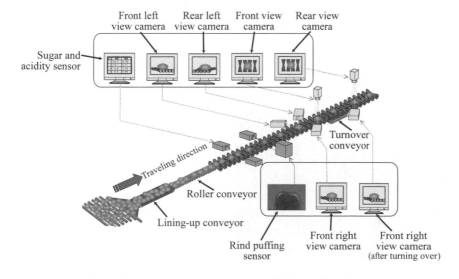

Fig. 3 *Example of inspection line (SI Seiko Co., Ltd.)*

At the inspection section, fruits are tested for internal quality criteria such as sugar content, acidity and rind puffing, then their appearance is checked by six color cameras taking images of the entire circumference of individual fruits as they are turned over 180 degrees by rotating roller pins (Figure 5). The fruits that pass inspection are carried to their predetermined grade positions and diverted to outgoing conveyors that are set perpendicularly to the main line by tilting of roller pins. Boxes of fruits are weighed by automatic weighing machines, shown in Figure 6, and marked with a barcode. This barcode is used later for the disclosure of product information and the printing of **grade and class** and weight information on boxes using inkjet printers. Boxes are sealed and loaded on pallets by palletizing robots and shipped. In this way, all work stages from receiving to shipping have been automated (Movie 3) and all the product information is accumulated in the host computer.

Preprocessing machines for spinach and spring onion[II-2.26] to cut off roots, remove outer leaves and peel outer skins, which were developed in 2001–2002,[16, 45] have been introduced in various regions. Lettuce packing machines which use arms for carrying and packing lettuces have also been put into practical use (Movie 4).

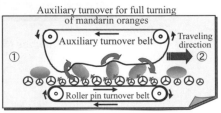

① Primary camera side   ② Secondary camera side

Fig. 4 Roller conveyor

Fig. 5 Mechanism of turning over by roller pins (SI Seiko Co., Ltd.)

Fig. 6 Automatic weighing machine

Forage crop harvesting involves side-delivery mowing, tedding, raking and picking. These tasks can be performed separately or in a single operation including mowing, chopping and blowing. There are various types of harvesters covering these tasks. Transport vehicles which follow harvesters are also likely to be automated or robotized first.

Forage crop harvesting is broadly divided into two types. In the harvesting system for hay and low to medium moisture silage, pasture grass stands are mowed by side-delivery **mowers** or **mower conditioners**, tedded and raked by **tedder rakes**, and windrowed forage is picked up and chopped up by **forage harvesters**, **hay balers** or **pickup wagons**. In the other harvesting system for high moisture silage and green forage, stands of pasture grass are harvested, chopped and blown by forage harvesters or **corn harvesters** in a single operation.

The configurations, functions and characteristics of various **forage crop harvesters** are explained below.[46, 47]

(1) **Mower**: Hair clipper-like **reciprocating mowers** and rotating blade type **rotary mowers** both perform side-delivery mowing. Rotary mowers are more popular, either as vertical axis type **disk mowers** (most common) and **drum mowers** or horizontal axis type **flail mowers** (also called hammer knife mowers). Many of them are attached to and powered by a tractor.

(2) **Mower conditioner**: The mower section performs side-delivery mowing and the conditioner section rakes in cut hay and performs conditioning work such as splitting and crimping. The conditioner section is either a flail type which splits and crimps hay with rotating flails or a crimper type which crushes hay between opposing rollers. They are typically either tractor mounted or tractor drawn (Figure 1), although some larger models are self propelled.

(3) **Tedder rake**: This is a combination of a **tedder** which spreads and turns over mowed hay and a **rake** which collects hay. Many types of tedder rakes are available depending on the structure and movement of the spring teeth (tine) section but the tedding and raking functions are not necessarily compatible.

(4) **Forage harvester**: Performs harvesting tasks from the mowing of forage crop or the picking up of windrows to chopping and blowing. Flail type harvesters are relatively compact. Unit type harvesters consist of a base unit which transfers, chops and blows forage crops and optional units for mowing and picking (Figure 2) (Movie 1). Many of the unit type forage harvesters are self-propelled and blow the crop into **transport vehicles** such as **forage wagons**. The corn harvester (maize harvester) is a specialized type of harvester for large crops such as green corn.

Fig. 1 Trailer type (PTO powered) mower   Fig. 2 Unit type forage harvester
conditioner (Deere & Company)

(5) **Hay baler**: An implement, sometimes simply called a **baler**, which picks up, compresses, forms and bales raked hay. There are **square balers** (tight balers) and **roll balers** depending on the shape of the bale they make. A baler consists of a pickup device for raked hay, a compress baling and binding device and a power transmission system. There are both tractor drawn and self-propelled types. One type of roll baler is a tractor drawn chopping type roll baler (Figure 3) which drives alongside a harvester, receives harvested and chopped material into a hopper and forms it into roll bales. To produce roll bale wrap silage, a machine called a bale wrapper is used to tightly wrap and seal bales (Movie 2).

(6) **Pick up wagon and forage wagon**: A pick up wagon picks up and chops raked hay and loads it on a tray for transport. It is also called a road wagon. A forage wagon drives alongside a forage harvester to collect and transport hay as it is harvested. It has a floor conveyor installed in its tray (so does the pick up wagon).

Many of the forage crop harvesting-related machines described are attached to tractors; therefore the robotization of this work requires automatic tractor driving technology. In joint operation by a harvester and a transport vehicle (Figure 4), automatic driving of the transport vehicle following the harvester is desirable for saving labor; research and development of **automatic following** technology is underway for this purpose.[II-4.6]

Fig. 3 Roll baler for chopped material   Fig. 4 Harvesting operation by forage
(Takakita Co., Ltd.)   harvester and forage wagon

Automated milking by robots is possible in a free-stall housing facility for dairy cattle. For tie-stall (stanchion) housing facilities where 80 percent of Japanese raw milk is produced, the commercialization of an 'automatic system for carrying milking units' developed by BRAIN and partners in 2002 is highly anticipated.

Significant labor saving in animal feeding and management is essential for 'dairy farming with latitude' (a more relaxed style of dairy farming) and the cost of such systems must be as low as possible. **Milking robots,**[II-3.23] which have been attracting attention in recent years, can perform milking operations unattended and thus offer significant labor saving. With this system, it is possible to milk cows twice or more a day and the amount of milk is expected to increase by 7–15 percent over the twice daily milking practice. There are two types of milking robots. In the parlor system, a single milking robot services two to four milking stalls and milker units. In the box system, one milking robot services one milking stall. The currently available parlor type milking robots include Liverty by Prorion of Holland, Zenith by Gascoigne Melotte of Holland, and Leonardo by Westfalia of Germany, and are capable of operating up to four milking stalls. The box type milking robots include Astronaut by Lely of Holland (Figure 1), Voluntary Milking System (VMS) by DeLaval of Sweden, Genesis by Bou-Matic of the U.S.A, and Dairy Dream by Orion Machinery of Japan.

In Japan, however, milk cows are fed at "**tie-stall** housings (stanchions)" in about 90 percent of the country's 30,000 dairy farms. Of these, 72 percent use pipeline milkers and 19 percent, mainly small operators, use bucket milkers. Combined, these farmers account for approximately 80 percent of Japan's total milk production. Consequently, it is highly desirable to develop low-cost technologies

Fig. 1 Milking robot (Astronaut Cornes AG, Lely)

Table 1 Specifications of automatic system for carrying milking units (BRAIN)

Milking procedures	Prototype model	Conventional method
① Carrying milking units	Automatic	Manual, push
② Connecting milk taps	Automatic	Manual
③ Washing teats	Manual	Manual
Attaching teat cups	Manual	Manual
④ Removing teat cups	Automatic	Manual, automatic
⑤ Dipping (sanitization)	Manual	Manual
⑥ Disconnecting milk taps	Automatic	Manual
⑦ Carrying milking units to the next cow	Automatic	Manual
Number of milking units carried	2	1
Milking at milk tap	2 cows simultaneously	2 cows alternately

to save labor and increase efficiency in the existing tie-stall setup. For this reason, BRAIN has developed an automatic system for carrying milking units jointly with manufacturers (Table, Figure 2).

The development project had two targets. First, one operator should be able to milk 50 cows per hour. Second, the cost of the system should not exceed ¥6,000,000. In order to achieve these targets, the following specifications were set for a prototype. First, the system automatically transports heavy milking units in pairs to two cows so that one operator can handle eight milking units whereas two to three operators used to be required to handle six units. Second, the procedures such as teat washing, teat cup attaching and teat sanitization are to be carried out manually since they require precision handling and are too costly to mechanize.

With the automatic system for carrying milking units operating suspended from overhead rails installed in the tie-stall housing, as shown in Figure 2, the following milking process occurs

1. The automatic system carries a pair of milking units to the designated pair of cows.
2. The system automatically connects to the milk vacuum pipeline and stands by.
3. The operator goes to the designated stall and cleans the teats of the two cows on either side and attaches cups.
4. The teat cups are automatically released when milking finishes.
5. The operator sanitizes the teats manually.
6. Once the completion of milking the two cows is confirmed, the system automatically disconnects from the milk vacuum pipeline.
7. The system automatically carries the milking units to the next pair of cows.

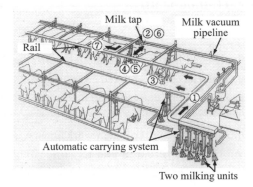

Fig. 2 Automatic system for carrying milking units (BRAIN)

Only teat washing, cup attaching and teat dipping need to be done manually; the rest is performed automatically.

The milking work rate (the number of cows milked per person per hour) using four automatic systems (eight milking units) is 52 to 56 cows, which is more than twice the work rate of 22 cows (with two to three operators using six units) prior to the introduction of the new system. The introduction of the automatic system for carrying milking units has established a technology to achieve labor saving and the milking of 50 cows per hour by a single operator in the tie-stall housing condition.

1.14 Carriers (power cart, monorail)

Carriers that are widely used in agriculture include self-propelled power carts for transporting farm materials and harvested crops in fields or facilities, trailers for transporting large machinery and hay, and monorails at inclined orchards.

Transportation accounts for a large part of farm work and a wide variety of carriers are used depending on the type of cargo and terrain.[48] Table 1 shows types of **transport vehicles** for various operational environments and purposes.

Self-propelled motorized power carts are widely used for transportation inside facilities or flat orchards. There are walking type power carts, riding type power carts and dual operation type power carts. The maximum loading capacity of a wheeled cart is around 150 kg for walking type one wheeler carts, around 250 kg for walking type three wheeler carts and about 500 kg for dual type carts. Four-wheeler transport vehicles are either dual operation type carriers or riding type carriers and many conform to small- or large-sized special vehicle standards and are permitted to drive on public roads. The maximum load capacity is around 600 kg in the case of small-sized special vehicles and several tons in the case of large-sized special vehicles. Some models are equipped with four-wheel drive and power steering (Figure 1). Crawler type vehicles fitted with rubber **crawlers** have superior

Table 1 Types of transport vehicles

Name		Diagram	Characteristics
Ackermann	Two-wheel steering		Most common. Also halftrack vehicles with tracked non-steering wheels.
	Single-wheel steering		Relatively simple structure. Small turning radius.
	Four-wheel steering		Complex structure. Tracking control is possible. Crab steering type can move in parallel (see diagram).
Articulated			Simple structure. Tracking control is possible. Suitable for a large fixed space.
Rotational difference (Caster)			Simple structure. Small turning radius. Poor running ability, not suitable for rough terrain.
Skid steering			Simple structure. Very small turning radius in pivot turn. Good running ability on rough terrain (it can damage road surface)
Ackermann tractor & trailer			Two-axle trailer is used for large/heavy load (see diagram). Trailer steering mechanism is somewhat complex.

Fig. 1 Power cart (small-sized special   Fig. 2 Power cart (crawler type)
         vehicle) (Atex Co., Ltd.)   (Atex Co., Ltd.)

grad-ability and turning ability and can be used on soft or inclined ground. Some models have a load capacity in excess of 1,000 kg (Figure 2). Additional features of power carts include carrying tray dump or lift functions, a driver's cabin for safety and comfort, and high-floor vehicles that can straddle crop rows.

Trailers tend to be tractor drawn and are used widely for transportation of large machinery and harvested forage crops as well as general cargos. Some models have additional features such as dump function and tray tilt function for the loading of machinery (Figure 3). In the area of forage crop harvesting, carriers such as **forage wagons**, **pick up wagons** and **bale wagons** are available.

For vertical transportation work such as loading of farm materials on and off a truck, implements and hoisting machinery such as tractor-attached **loaders**, **skid steering** loaders and **forklifts** are used. These can be regarded as transport vehicles in a broad sense.

Monorail carriers are commonly used for transportation at steeply inclined orchards (Figure 4).[49] A monorail carrier is a combination of a traction engine

Fig. 3 Trailer (Star Farm Machinery   Fig. 4 Monorail (Nikkari Co., Ltd.)
         Mfg. Co., Ltd.)

and a towed cargo carrier which travels on a fixed track and generally uses the rack and pinion system by which the pinion type drive wheel travels along the rack type rail. Since it climbs up and down slopes as steep as 50 degrees, it is usually equipped with three brakes for extra safety, including the descent brake to maintain constant descending speed, the park brake to stop and hold the vehicle, and the emergency brake which activates when a designated speed is exceeded. These improved safety features have facilitated the use of monorails for carrying passengers and work vehicles. There are multipurpose models that can be used for tasks such as chemical and fertilizer application and tree pruning as well as transportation.[II-4.18] (Movie)

Coffee break

Should agri-robots 'look' good too!?

Robotics has come a long way from the production of industrial robots in the twentieth century to the development of robots that are more relevant to the everyday lives of ordinary people, including biped robots such as ASIMO, building management robots, home security robots and even mental commitment robots for therapeutic purposes which have been developed and publicized in recent years. The development of robots for hazardous jobs at heights, such as the inspection and painting of the external walls of tall buildings, and disaster management robots is also underway. Robots are becoming indispensable for the creation of a safe and secure society.

In agriculture, the development of workload reducing, labor saving technologies such as robots is an urgent task and addressed in the Agriculture, Forestry and Fisheries Research Basic Plan launched by the Ministry of Agriculture, Forestry and Fisheries on 30 March 2005. Advances in this direction is essential for stable production and supply of food in Japan. While there are high expectations for robotics as Japan's next generation food production technology, we must not neglect to assess their economic efficiency. We must consider cultivation conditions that are more suitable for robotization. In other words, the robotization of food production requires designing a regional environment for harmonious interface with robots as well as designing new cultivation systems in collaboration with thremmatology and agronomics. There are still countless issues to be resolved. For example, my lab often conducts a public viewing of robot tractors it has developed. We receive some harsh comments from farmers and the general public such as 'the robot looks spooky'. Perhaps the sight of an ordinary tractor with handlebars traveling and working in a farm unmanned may look odd. In an ideal picture of robot farming, robots would blend into the rural environment in the hills and mountains of Japan. (N. Noguchi)

2 Agri-Vision

Due to the diversity and complexity of its target objects, machine vision technology for agricultural robots entails various conversion methods, computation methods, and algorithms for recognition, identification and position detection. As image input devices, TV cameras with photographic sensitivity ranging from X-ray to infrared are used, with a strong indication that the terahertz region will be in use in the near future. This chapter calls the machine vision of agricultural robots 'agri-vision' and explains the types of hardware and software that are suitable for the optical and physical characteristics of various target objects.

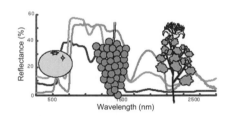

2.1 Underground sensing (Movies)

An optical soil sensor capable of real-time measurement of the moisture, organic matter, nitrogen, EC and pH levels in the plow layer was developed by a team led by Tokyo University of Agriculture and Technology. It was turned into a working prototype by Shibuya Machinery Co., Ltd. in 2004. The sensor has made the recording and processing of underground images possible.

Historically, farmers have tended to apply the same types and amounts of fertilizer every year based on their experience. However, soil analyses of fields have found that this frequently causes serious problems as a result of chemical fertilizer use. Tokyo University of Agricultural Technology has been developing and refining a real-time **soil sensor**[1] that attaches to a tractor's **three-point linkage hitch system** as part of a project to promote community-based precision agriculture in Japan.

Figure 1 shows the structure of the soil sensor and Figure 2 shows the soil sensor attached to a tractor. This sensor travels at a speed of 1 km/h as its chisel bores a flat-based tunnel in the soil at a depth of 10–40 cm (Movie 1). It transmits light from a halogen lamp to the sensor casing through **optical fiber** cables and illuminates the base surface of the tunnel. The light reflected by the flat tunnel base is sent to the measuring equipment on the ground through two optic fiber cables. The received light (400–1700 nm) is then separated into 256 channels in the **visible region**[1-2.1] and 128 channels in the **near infrared region.**[1-2.1] Appropriate wavelengths are then used for measuring the moisture, organic matter, nitrogen, **EC** and **pH** levels of the plow layer soil. When the position of the sensor within a field is measured by **GPS,**[1-4.1] these measurement data can be turned into maps[2] as shown in Figure 3 which can be used in long- and short-term field management taking into account these data. At the same time, images of the tunnel base can be recorded with a small color **TV camera**[1-2.3] located between the sending and receiving optical fiber cables (Movie 2).

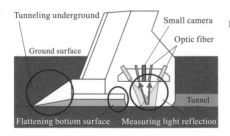

Fig. 1 Structure of soil sensor[2]

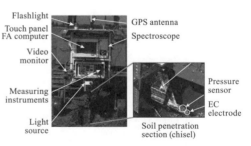

Fig. 2 Soil sensor attached to tractor

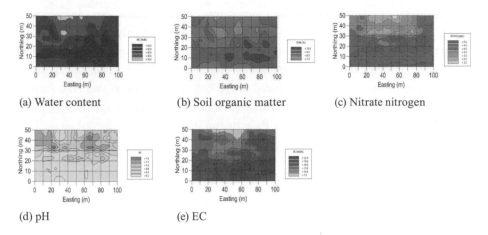

(a) Water content   (b) Soil organic matter   (c) Nitrate nitrogen

(d) pH   (e) EC

Fig. 3 Field maps by soil sensor

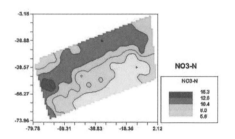

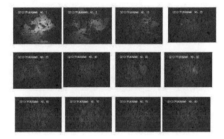

Fig. 4 Example of nitrate   Fig. 5 Underground soil images
nitrogen-N map   *(Fig 1–Fig 5: Tokyo University of*
Agriculture and Technology)

Figure 4 is an example of a map showing the distribution of **nitrate nitrogen** in a field where the soil sensor was run at a depth of 15 cm on 17 October 2002. This map was prepared based on data from a total of 436 points gathered at intervals of 1 m. The sample field is a large-scale paddy field in Fukagawa, Hokkaido, created about eight years ago. The right side of this field had low levels of nitrates and suffered the lodging of rice plants every year despite a low level of fertilizer. Figure 5 shows images taken by the TV camera of the soil sensor, which was run at a depth of 30 cm. The four images on the first row and the left two images on the second row were recorded at intervals of 5 m or 10 m starting from the left end of the field. The last image at the lower right was taken at 80 m from the left end, i.e., close to the right end of the field. It was possible to surmise based on these images that rice lodging was caused by the existence of plowed soil at a depth of 30 cm on the right side of the field whereas yellow subsoil was visible at this depth on the left end of the field.

Additional research[3] is being conducted into ways to extract soil information that cannot be obtained by near infrared spectroscopy by analyzing the color and **texture** of the soil in underground images. The development of software for real-time processing of such images is also underway. About 130 soil sensors capable of measuring EC have already been introduced in the U.S. These sensors can take various measurements and their ability to directly gather field information is receiving much interest. Similar research projects[4, 5] have begun in the U.S. and Europe as well.

Chrysanthemum cutting sticking

> This machine vision system used on a robot for automated chrysanthemum seedling production was developed by Okayama University, JA Aichi and Matsushita Electric Industrial Co., Ltd. from research which began in 1995. The first unit was introduced to the farming business support center of JA Aichi.

In this system, chrysanthemum stems stored in a refrigerator are fed into a cutting feeder unit which separates entangled stems in a water trough into individual cuttings. They are transported on a conveyor belt as the vision system shown in Figure 1 recognizes the orientation of the cuttings. The end-effector of the **cutting sticking robot**[II-3.2] then grabs the end of the main stem (Movie).

The vision system of the cutting sticking robot[6–9] has a camera installed above the conveyor belt to capture images of the cuttings. Image signals from the camera are entered into a computer with a capture board so that the grasping position for the end-effector is recognized on the computer. At this stage, the conveyor belt may be contaminated with water drops and dirt. The system uses a wavelength of 850 nm to ensure high reflectance from the stem cuttings in order to easily separate these noises from images of the cuttings in the image processing. For this purpose, it has a filter with a transmission range of 850 nm ± 5 nm in front of the black and white camera.

Once the image is captured, the main stem is recognized as follows. First, the **contour** of the binarized image[I-2.4] is traced. The orientation of the pixels next to the focal pixel is recorded using a value called **chain code**. This method allows the outline of the image to be described using only the chain code so that the contour line can be reproduced. Figure 2 shows an example of chain coding.

Before moving on to an explanation of the algorithm for determining the endpoint of the main stem, a method to quantify linearity from the shape of a contour should be explained. The linearity in question here is evaluated by distance L as shown in Figure 3. Each of the thick sections of the outline has a standard length (yardstick) of fifteen pixels before and after the focal pixel which is used to ascertain the value of L. In this example, the relationship between three Ls ($L_c < L_a < L_b$) suggests that the smallest L value of L_c has a high linearity and therefore resembles the shape of a stem.

An algorithm for main stem detection is then applied using a **binary image**[I-2.4] of the cutting taken by this system and a graph with a linear distance value L of each outline pixel on the vertical axis. As per the binary image on the right of Figure 4, an image of an actual cutting may have more than one segment that resembles the shape of the main stem. Humans are able to identify the main stem

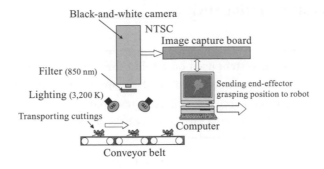

Fig. 1 Machine vision system of cutting sticking robot

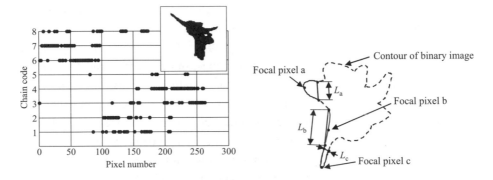

Fig. 2 Example of chain coding   *Fig. 3 Quantification of linearity*

correctly by utilizing information other than linearity. It is likely that humans subconsciously take spatial distribution of the leaves into consideration.

The graph on the left side of Figure 4 shows the linear distance values of this image. Candidates for the possible end-point below a threshold of T_1 include points A, B and C. When the linearity around the candidate points are evaluated with regard to a threshold of T_2, point D is found to be below T_2 and therefore point A is found to have a nearby segment which disturbs the linearity. Accordingly, point A is excluded as a possible end-point and either point B or point C is considered to be the end-point of the main stem. At the same time, likely leaf tips are identified using a threshold of T_3. Points D, E and F are recognized as leaf tips and, together with point A which was excluded earlier, presumed to be leaves.

Straight lines connecting the gravity center (point O) of the binary image with leaves and possible stem end-points, as shown in Figure 5, are used in determining whether point C or point B is the end-point of the main stem. The measuring of an angle (θ) formed by two adjacent points of point C and point B respectively centering on point O indicates $\theta_c < \theta_b$. It is determined that point C is located

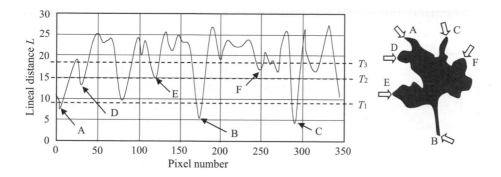

Fig. 4 Extraction of candidate points

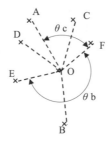

*Fig. 5 Gravity centre of cutting
and candidate points*

within the leaf region and that point B, which is located opposite to the leaf region, is the end-point of the main stem.

Based on this analysis, the positional information is transmitted to the cutting sticking robot, which grasps the cutting with its end-effector and automatically proceeds to stick the cutting.

2.3 Carnation and geranium cutting sticking　　　　　　　　(Movie)

> A carnation cutting sticking robot and a geranium cutting sticking robot were introduced in 1988 by Toshiba Corporation, Japan, and University of Georgia, U.S.A., respectively. The latter was developed in association with the Georgia Agricultural Experiment Station and the NASA Marshall Space Flight Center.

Carnations are often reproduced by vegetative propagation using stem cuttings with auxiliary buds. The work has traditionally been done manually but the automation and robotization of this procedure is beneficial in terms of preventing viral contamination as well as saving labor. Figure 1 shows the seedling production system. This propagation robot system propagates a carnation plant by recognizing the nodes of a seedling in a tray, cutting it into individual segments from the top, and sticking them in a new medium in a tray one by one. It was exhibited at the 1990 International Garden and Greenery Exposition.[10] This system consists of a recognition robot (three-degrees-of-freedom **Cartesian coordinate type**[(I-3.6)]) that detects and recognizes seedlings and a transplanting robot (six-degrees-of-freedom **articulated type**[(I-3.6)]) that cuts stem segments and sticks them in a new medium. The recognition robot (Figure 2) is fitted with a **triangulation**[(I-2.4)] type range sensor, not a TV camera,[(I-2.3)] which uses a near infrared **laser beam** to scan seedlings and a position sensitive device (**PSD**)[(I-3.4)] to receive reflected light via a lens (Figure 3). Laser beam scanning is done horizontally from the lower part to the upper part of the seedling using a **galvano mirror** and the 3-D shape of the whole seedling is sent to a computer. Based on this 3-D shape, the computer identifies branched segments as nodes, determines the positions for grasping and cutting by the transplanting robot, and issues a command to the transplanting robot (Figure 4). At the 1990 Expo, the time it took from seedling recognition to transplanting was approximately 10 seconds.[10]

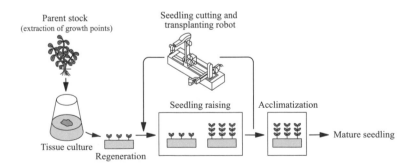

Fig. 1　Carnation seedling production procedure[10]

56

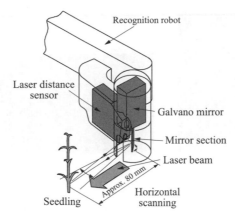

Fig. 2 Seedling recognition robot[10]

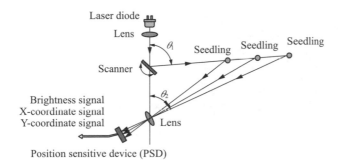

Fig. 3 Principle of position detection[10]

Geraniums are also propagated by stem cuttings as shown in Figure 5. A robot system[11] shown in Figure 6 has been developed for this purpose. It first detects geranium cuttings laid on the conveyor belt using the TV camera installed above. It uses the image to recognize nodes and stems and to determine grasping and cutting positions. It then uses an end-effector to grasp the cutting and uses a **polar coordinate arm**[1-3.6] to carry the seedling to the lower leaf remover. After identifying its shape by the infrared **array sensor** of the bend detector, the robot sticks the seedling in a special plug tray (Movie).

In image processing,[12] the system recognizes the leaf stem branching structure of the seedling and identifies the position and orientation of the seedling and the junctions of the main stem and leaf stems (Figure 7). It also determines whether or not there is sufficient space for the left and right fingers between leaves and leaf stems in the vicinity of the central grasping position of the main stem for ease of

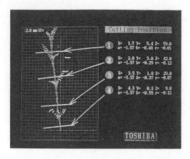

*Fig. 4 Detection of seedling
cutting positions*[10]

Fig. 5 Geranium seedling

*Fig. 6 Geranium cutting
sticking robot*

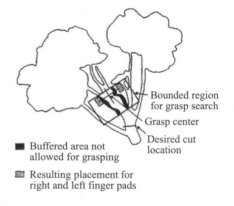

*Fig. 7 Recognized grasping
and cutting positions*[12]

grasping by the end-effector. It issues instructions as to the position for cutting off
the unnecessary part of the lower stem and lower leaves. A trial run on over 400
stem cuttings produced no grasping failure by the end-effector and only minor leaf
stem damage to 1.5 percent of the cuttings. The process was completed within an
average time of six seconds or so per cutting (Figure 8).[12]

*Fig. 8 Cutting sticking
by robot*

2.4 **Crop space weeding** (Movie)

A joint research project by Shimane University and the Hyogo Prefecture Agricultural Research Center began in 1991 to develop a crop space weeding robot for initial weeding in the transplanting cultivation of leaf vegetables. Crop space weeding was studied as one of the tasks to be completed by a multi-functional robot for vegetable production. Research was conducted into the vegetable transplanting and harvesting functions of the same robot.

Weeding can be done by hand, using a harrow or **cultivator**, burying in the ground by harrowing, or chemical control with herbicides. Unlike inter-row weeding along crop rows, it is difficult to apply traditional mechanization technologies to intra-row weeding as the weeder must steer clear of crops. Accordingly, research was undertaken on crop space weeding by robot for the purpose of automating the removal of weeds that are interspersed between crops (Figure 1) (Movie).[13] Crop space weeding was studied as part of research and development of a **multi-functional robot** for vegetable production capable of transplanting and harvesting operations as well. The robot has detachable end-effectors for different operations.

This research project into crop space weeding targeted the initial weeding in the transplanting cultivation of leaf vegetables. The robot uses the **stereo vision**[1-2.4] of a color video camera to detect 3-D positions of weeds and extracts weeds with some surrounding soil using spiral-shaped rotary blades attached to an end-effector. The traveling section is a four-wheel-drive vehicle that straddles the ridge. The guide rollers fitted to the front wheels contact the side slopes of the ridge to guide the robot along. In most cases, trials have been conducted using commercial power sources but it is possible to operate the machine by mounting a small generator of 300 W or so.

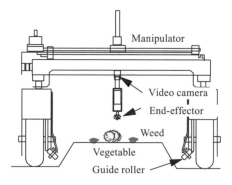

Fig. 1 Weeding between crops by a robot

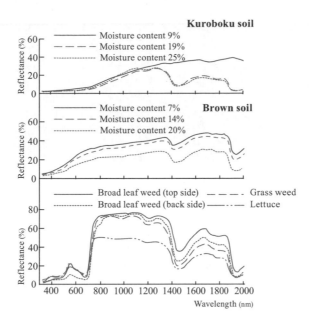

Fig. 2 Spectral reflectance of soil and plants

Crops and weeds tend to exhibit similar patterns of reflectance as shown in Figure 2 which peak at a wavelength of around 550 nm in the green range, dip at around 670 nm in the red range, and rapidly rise at around 800 nm in the **near infrared region**.[1-2.1] In contrast, soil reflectance varies according to the moisture content and does not have a peak in the green wavelength range nor a dip in the red wavelength range. It is easy to distinguish the soil from plant bodies due to this clear color difference but not so easy to distinguish between weeds and crops due to their similar coloring. Consequently, the image processing program looks for **binary images**[1-2.4] of plant bodies, including weeds and crops. In the case of transplanting cultivation of leaf vegetables, weeds and crops are distinguished based on their sizes in binary images since seed-germinated weeds are smaller than transplanted crops.

A color image in which one pixel represents eight bits each of the red, green and blue colors are entered into the computer and the image processing program produces a binary image which recognizes plant bodies based on pixels having a green component exceeding the average quantities of blue and red by a certain range. When the image of weeds (*Erigeron annuus*) in Figure 3 is converted by the binarization[1-2.4] process, plant bodies appear as white in a binary image in Figure 4. Weeds and crops are distinguished based on their sizes using the horizontal and vertical **Feret's diameters**[1-2.4] obtained from the binary image.

 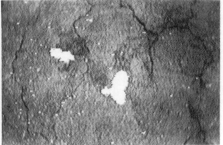

Fig. 3 Original image of weeds   *Fig. 4 Binary image of weeds*

The traveling section is stationary while an image is being acquired. A downward-pointing camera is attached to the **Cartesian coordinate arm**.[1-3,6] The arm carries the camera for 50 mm in parallel to the ridge to take one image each before and after the move. The stereo vision system calculates the 3-D positions of weeds based on these two images. Stereo matching is based on three criteria: When the two binary images are compared,

(1) image shear in the directions other than the direction of camera movement is small;

(2) image shear in the direction of camera movement is within a certain range; and

(3) the horizontal and vertical Feret's diameters in the two images are about the same.

It has been reported that the 3-D positions of weeds could be measured by the image processing program with minor errors up to a few milimeters.[13]

2.5 Crop row recognition

> Crop row recognition capability provides a basis for automatic vehicle guidance in field management and automatic operation of unmanned vehicles such as tractors. A side sliding actuator fitted between the implement and the three-point linkage hitch system can control the positioning of the cultivator and make cultivating and weeding operations easier since the driver does not have to pay attention to the positioning of the implement while driving.

Various sensing systems have been proposed for crop row recognition. The most common method among them is **machine vision**.[1-2.1] This recognition method[1-4.2] distinguishes crops from the soil and approximates a crop row with a straight line by the **Hough transformation**[1-4.2] or the regression line on the basis that a crop row is generally straight. It aims to detect lateral deviation and directional deviation of the vehicle or the implement from the center of the crop row in an image coordinate system or a vehicle coordinate system.

Commercial RGB video cameras or monochrome cameras fitted with a near infrared interference filter are used as machine vision hardware. A critical issue for any of these sensors is the capability to discriminate the crop region and the soil region consistently and accurately.

With machine vision, when lateral deviation and directional deviation are detected in a 2-D image coordinate system, at least lateral deviation is expressed in number of pixels and detected values are affected by the specifications of the vision sensor, therefore it generally needs to be converted to a vehicle coordinate system. The explanation of the conversion method is omitted here

(a) Original image   (b) Conversion to vehicle coordinate

Fig. 1 Coordinate conversion of crop row image (Hokkaido University)

since it is described in Section 4.2.5 of *Agri-Robot (I)*. The original image in Figure 1(a) looks like the image of crop rows taken from above in Figure 1(b) after coordinate conversion. It is called vehicle coordinate because it is a plane, with the orientation of the vehicle as the coordinate axis. The slope of the crop row line is the directional deviation of the vehicle and the lateral deviation of the crop row can be detected at arbitrary positions in front of the vehicle.

Crop row detection by binocular **stereo vision**[1-2.4] has been studied in recent years. Since the sun is the light source in the outdoor environment, **image segmentation** based only on color information is considered to be inadequate due to color temperature variations. Also, it is difficult to distinguish crops from weeds based only on color information. However, stereo vision is capable of stable image segmentation between crops, weeds and soil because it can provide distance information as well as color information and use the size of vegetation as a feature quantity.

Figure 2 shows a stereo vision system fitted to a **tractor**. Figure 3(a) is the image of crop rows taken by the stereo vision system. Stereo vision can provide the **distance image**[1-4.2] (Figure 3(b)) in addition to the luminance image. Since the distance between the camera and crop rows is shorter than the distance between the camera and the soil, image segmentation based on distance information is possible. In other words, both the precision and robustness of crop row detection can be enhanced by hybridization of the luminance image and the distance image.

Thus, the information from crop row detection can be used to control the position of both the vehicle (Figure 4) and the implement (Figure 5) (Movie).

Fig. 2 Stereo vision system mounted on tractor

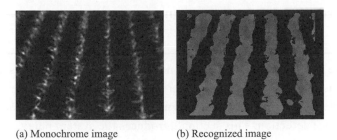

(a) Monochrome image   (b) Recognized image

Fig. 3 Crop row recognition by stereo vision

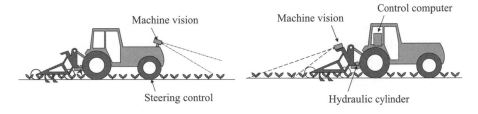

*Fig. 4 Autonomous traveling by
machine vision (Hokkaido
University)*

*Fig. 5 Implement control by
machine vision (Hokkaido
University)*

2.6 External recognition by laser scanner (Movie)

> While the TV camera is the most common external sensor for robots, external sensing using lasers is becoming increasingly common. Laser sensing has many advantages, including faster and more accurate object shape recognition and distance detection. This section introduces the system being researched at Okayama University.

Various range finders are currently available on the market, some of which can be connected to a personal computer for easy measurement and control. One such product is the **laser scanner** (by SICK, Inc.)[14] [I-2.5] shown in Figure 1. An internal spinning mirror shoots a laser beam at objects in variable directions and the machine detects distance based on the time the reflected laser beam takes to come back (time-of-flight). In other words, it detects surrounding objects two-dimensionally (in fan shape). The maximum scan angle is 180 degrees and the scan interval can be set at 0.25, 0.5 or 1 degree. Gathered distance and angle data can be fed into a computer through RS-232 or RS-422 interface. It uses a Class 1 eye safe type infrared laser diode as the light source, and can therefore be used for detecting humans. Outdoor models are available with functions that compensate for the effects of fog, rain and snow.

The target objects of agricultural robots have different forms or different objects depending on the height level in many cases. For example, while a support pole for a grapevine trellis stands almost vertically and maintains the same form at most height levels, the trunk of a grapevine has gnarls, bends and branches. Bunches of grapes and leaves hang from the trellis. Consequently, the robot requires 3-D information in order to avoid obstacles and complete its tasks. When the laser scan in Figure 1 is fixed, it scans within a 2-D plane parallel to the ground. When vertical movement is added, it can obtain 3-D distance information.

Figure 2 shows a laser scanner mounted on a vertical travel device.[15] The table moves up and down for a stroke length of 300 mm by a motor and ball screw drive and its positions are recorded by a **rotary encoder**.[I-3.4] The laser scanner descends a certain distance between each scan to gather 3-D distance information (**3-D image**). In this example, the scanner moves around on a travel device in a tomato-growing greenhouse and scans individual plants. If it needs to scan at a place higher than the stroke of the vertical travel device, the whole device is elevated by a lift.

Figure 3 shows a sensing system[15, 16] mounted on a **robot to work in a vineyard**. [II-3.5-7] This robot determines the movement of its arm to evade obstacles based on 3-D images from the laser scanner and constantly monitors the position and speed of humans in the vicinity to ensure their safety (Movie).

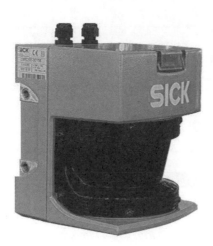

Fig. 1 Laser scanner

*Fig. 2 Laser scanner mounted on
vertical motion device*

Figure 4 shows an example of 3-D distance information scanned in a vineyard. Figure 4(b) is the 3-D image obtained by the scanning of the boxed section of the view of the vineyard in Figure 4(a). The machine scanned 401 times at 0.25 degree intervals horizontally and 100 times at 3 mm intervals vertically. In this example, each detection point is described in gray scale according to its distance for ease of understanding (Figure 4(c)). The branched part of a tree, the shape of leaf and the support pole are clearly detected. Leaves and branches behind the pole were excluded from the 3-D image since the detection range was set at 2.5 m in this example. The range finder has the advantage of being able to gather information within a required range while the **vision sensor** gathers all information within its field of vision.

(a) Original image

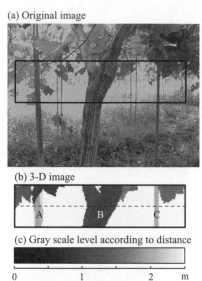

(b) 3-D image

(c) Gray scale level according to distance

0   1   2   m

Fig. 3 Sensing devices mounted on robot to work in vineyard

Fig. 4 3-D image

2.7 Tomato and cherry tomato harvesting

Research and development of a tomato harvesting robot began at Kyoto University in early 1980s which sought to position fruits in 3D using stereo vision. Subsequent research at Shimane University and Osaka Prefecture University seeks to develop recognition systems for foliage and individual fruits in a bunch or cluster for the harvesting of fruits while avoiding obstacles using 3-D vision sensors.

The **3-D vision sensor**[I-2.5] being used in research on the tomato and cherry tomato harvesting robot[II-3.11–12] scans fruits using a **laser beam** having two different wavelengths to measure the **3-D shape** as well as to distinguish foliage and immature fruit from red-ripe fruit. Figure 1 depicts the optical system. It uses a cold filter to overlay a near infrared beam and a red laser beam. The light reflected by an object passes through a lens and forms an image on the photosensitive surface of the **PSD** (position sensitive device).[I-3.4] Distance is calculated by **triangulation** based on current ratio between two anodes. The emitted laser beam is blinking and the receiver unit extracts only the blinking wavelength components, distinguishing them from the fixed light of the sun.

When a laser beam with two different wavelengths is emitted, the PSD receives the light components of both wavelengths. The near infrared and red laser beams blink at different frequencies so that demodulator circuits can extract their different frequencies. As shown in Figure 2, foliage and fruits all reflect the near infrared laser beam well but forage and immature fruits have a low red laser beam reflectance. Red fruits are detected by the 3-D image processing program based on this principle.

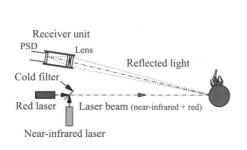

Fig. 1 Optical system of three dimensional vision sensor using two-wavelength laser beam (Shimane University)

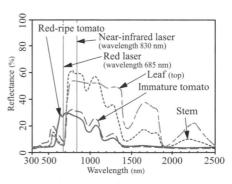

Fig. 2 Spectral reflectance of tomato plant

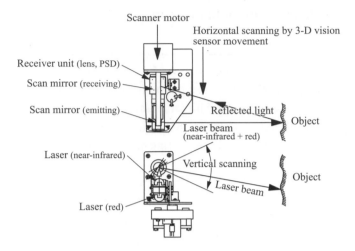

Fig. 3 3-D vision sensor (Osaka Prefecture University)

Figure 3 shows an example of the 3-D vision sensor[17] which scans with near infrared and red laser beams. This unit is being used in research for a **multifunctional robot** for tomato cultivation which trains vines and defoliates as well as harvesting. The 3-D vision sensor is attached to the end of the robot's arm, performing vertical scanning by swinging a scan mirror using a stepping motor[1-3.3] and horizontal scanning by horizontal movement of the arm. The sensor is capable only of vertical scanning on its own. Since it can obtain voltage values of the near infrared ray and red light for individual pixels, it can recognize red fruits, green foliage and immature fruits, main stems and support posts individually using the image processing program. Consequently, it is suitable as a vision sensor for the harvesting of fruits while avoiding obstacles such as foliage and support posts. It is also being trialed as a vision unit for vine training and defoliation due to its ability to recognize individual parts of a plant body.

A 3-D vision sensor which scans using a two-wavelength laser beam based on the same principle is also used in research seeking to develop a **cherry tomato harvesting robot**.[18] Cherry tomato fruits grow in clusters as shown in Figure 4. The left image in Figure 5 is a representation of pixels of the objects located within a certain range derived from an image obtained by the 3-D vision sensor. Based on voltage differentials between the near infrared and red light signals, pixels with greater red values are shown as white and all others are shown in gray. Although white training twine as well as red-ripe fruits appear white, 3-D image processing can extract fruits only. Based on the fact that the fruit has a glossy surface and the light voltage peaks when the center of the fruit is scanned, the sensor can recognize clustered fruits as long as most of them are visible. The right

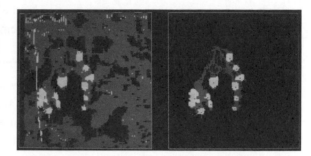

*Fig. 4 Cherry tomato
    fruit cluster*

Fig. 5 Results of image processing

image in Figure 5 shows extracted fruits as white and obstacles around them in gray. The 3-D vision sensor used in this experiment simultaneously scans three points by splitting a laser beam of near infrared and red wavelengths into three beams, scanning 120 x 120 pixels in approximately two seconds.

2.8 Cherry tomato and strawberry harvesting (Movie)

> Visual feedback processes are used to recognize fruit growing in clusters such as cherry tomatoes and strawberries by the machine vision systems of harvesting robots. This visual feedback system was trialed at Okayama University from 1995 to 1999.

Cherry tomatoes and strawberries grow in clusters as shown in Figure 1. Fruits are usually harvested one by one as they mature although cherry tomatoes are sometimes harvested still attached to a truss. A **visual feedback** system using an ordinary color camera was researched at Okayama University to provide **machine vision**[I-2.1] for robots for harvesting these crops. Figure 2 shows the machine vision system. A color **TV camera**[I-2.3] and an **image processing board**, both **VGA**[I-2.1]-class, and an **incandescent lamp**[I-2.3] with a color temperature of 5500 K are used. The camera is moved laterally by 150 mm, taking one image each before and after the move in order to obtain **stereo vision**.[I-2.4] It can also move vertically by a distance of 400 mm. Combined with the vertical mobility of the robot itself, the camera can cover a visual field between approximately 400 mm and 1600 mm above ground at a distance of 900 mm from the camera.

Figure 3 shows an outline of the control process for the harvesting robot.[II-3.10] First, a binary image[I-2.4] is obtained by image operation performed on (Y–R) and (R–G) in order to extract ripe fruit only. R and G denote red and green images and Y denotes a luminance image calculated by formula 2-13 in Section 2.4 of *Agri-Robot (I)*. It is generally difficult to match two stereo images of cherry tomatoes, which sets many small fruits in a single cluster, and false matching is common. Accordingly, distance data for the center point of a polygon is adopted as a reference value and distance data for each apex of the polygon are compared. Those with small differences are regarded as reliable data and used in the estimation of the center position of the cluster figure. In the example shown in Figure 1, harvesting action was performed by moving the robot arm toward a fruit in the upper left position which was most ripe in the cluster and posing a very low risk for the end-effector to contact adjacent fruits. When the presence of the fruit was detected by a **photo-interrupter**[II-3.10] attached to the end-effector, difference between the detected position and the position calculated from image information was checked. When a re-input image confirmed that the target fruit had been duly picked, the system moved on to compute the position of the next target. This process of image input and harvesting action was performed repeatedly (Movie).

Fig. 1 Cherry tomato and strawberry fruit clusters

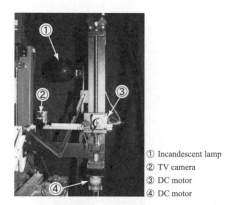

① Incandescent lamp
② TV camera
③ DC motor
④ DC motor

Fig. 2 Machine vision system

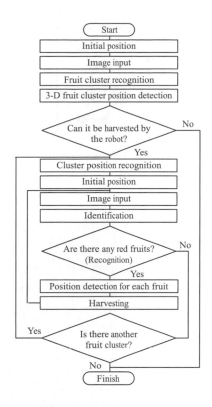

Fig. 3 Flowchart of robot control

Figure 4(a) shows the fruits detected from the input image of the tomato cluster in the left image in Figure 1. Only five red fruits were detected and the distance to the fruit cluster computed by stereo vision was 690 mm. After fruit A was picked (Figure 4(b)), the cluster position changed slightly and the orange-colored fruit F was newly detected. When fruits B and C were picked, fruit G behind them was detected (Figure 4(c)) together with fruits D, E and F. This harvesting action was repeated until there was no more ripe fruit left. In the end, all the fruits were harvested as shown in Figure 4(d).[19, 20]

Stereo vision is also used for strawberry harvesting. Due to difficulty in image matching as shown in Figure 5, however, the following approach is being tested. In **table top culture**,[(1-3.3)] the distance between the robot and the fruits is largely determined by the distance between the robot and the table. The robot harvests the first fruit based on an approximate distance given as the initial value, then computes the 3-D position of the next fruit based on the coordinates of the harvested fruit.[21] Different combinations of these approaches are being considered for practical application.

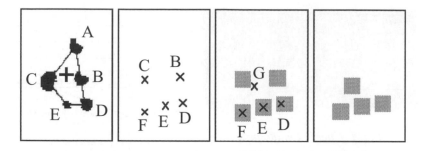

Fig. 4 Results of fruit detection

Fig. 5 Images acquired by
stereo vision

2.9 Eggplant harvesting

Research and development of a robot for eggplant harvesting began in 1997 at the National Institute of Vegetable and Tea Science. Color CCD cameras were used for visual sensing and a basic algorithm for eggplant fruit recognition based on color information and morphological characteristics was developed.

Eggplant fruit, stem and leafstalk present a deep aubergine color (dark purple) and can be distinguished from green leaves based on color information. Although spectral reflectance[1-2.2] analysis of each part of the plant shows that the lignified (woody) stem has a slightly higher reflectance than the fruit at around 500–700 nm, however, this color information is not sufficient to discriminate the fruit from the stem. Consequently, a **fruit recognition algorithm** was developed using shape differences between the stem, the leaf stalk and the fruit as well as color information.

The visual sensing unit of the **eggplant harvesting robot**[II-3.13] has a **global sensing** function to detect fruits at the initial arm position and a **local sensing** function to determine the harvesting suitability of the fruits once the arm has moved closer to them.[22, 23] A CCD camera[1-2.1] is used as a vision sensor, fitted at the center of the end-effector so that a single vision sensor can perform both global sensing and local sensing to acquire images (Figure 1). The image processing algorithm for global sensing is based on color information and morphological characteristics.

First, an RGB image is processed for extraction of non-green segments by **binarization**[1-2.4] of a G–B grayscale image. Next, a G image is binarized for the extraction of non-background segments. Since the level of illumination at the time

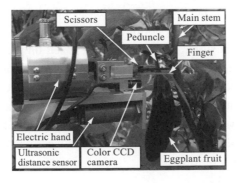

Fig. 1 Vision sensor attached to end-effector

75

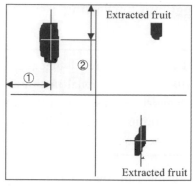

① Center of gravity coordinate F_{H}
② Center of gravity coordinate F_{V}

Fig. 2 Acquired image of eggplants   *Fig. 3 Extracted image of fruits by*
                                                              global sensing

of sensing and the amount of foliage vary, discrimination analysis is used for the determination of **threshold values** for binarization. The **logical product** of the two images is obtained to extract dark purple segments and **erosion**, **dilation** and **filling**[1-2.4] are performed to enhance the shape. By eroding and dilating five pixels horizontally and twenty five pixels vertically, any part of the target object 500 mm in front of the sensor that is about 11 mm wide or less (fruit stalks and stems) or 55 mm long or less (small fruits not yet suitable for harvesting) is excluded. Since two to three nodes are included in an image segment and each node sets up to two harvesting-stage fruits, detection of a maximum of five fruits within an image segment means no omission of harvesting-stage fruits. Up to five larger objects among the eroded and dilated objects on the image are recognized as fruits and the area and barycentric position of each fruit are calculated. Figure 2 shows the acquired image of eggplants. Figure 3 shows the extracted image of the fruits in global sensing.

Next, local sensing is performed once the tip of the arm has approached a fruit as part of the robot's harvesting action in order to estimate the position and size of the fruit and the position of the fruit stalk based on the image data and the distance data from an **ultrasonic sensor**.[1-3.4] The image processing algorithm for local sensing is largely the same as the one for global sensing which detects fruits based on color information and morphological characteristics. However, the horizontal and vertical erosion and dilation S_{H} and S_{V} are calculated using the following formulae based on the distance between the camera and the fruit measured by the ultrasonic sensor.

$$S_{\mathrm{H}} = \frac{R_{\mathrm{H}} L_{\mathrm{H}} \tan(a_{\mathrm{H}}/2)}{2D} \qquad\qquad (1)$$

$$S_{V} = \frac{R_{V} L_{V} \tan(a_{V}/2)}{2D} \qquad (2)$$

a_{H} and a_{V} denote the camera's horizontal and vertical angles of view (°) respectively, R_{H} and R_{V} denote the horizontal and vertical **resolutions** (pixels) of the image, L_{H} and L_{V} denote the horizontal and vertical lengths (mm) of the object for exclusion, and D denote the distance (mm) from the camera to the fruit.

After excluding any part that is about 12 mm long horizontally and about 50 mm long vertically by erosion and dilation, the largest part of the detected object is singled out and its center of gravity coordinate and maximum length are obtained as shown in Figure 4. Of the two points used to determine the maximum length, the one closer to the upper end is regarded as the stalk end of the fruit and it is assumed that there is a fruit stalk above this point. Finally, since the fruit is hanging down almost vertically, fruit length L_{F} (mm) is estimated using the following formula modified from formula (2) above. L_{MAX} is the maximum length (pixels) of the detected object.

$$L_{F} = \frac{2L_{H} D \tan(a_{V}/2)}{R_{H}} \qquad (3)$$

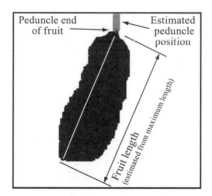

Fig. 4 Extracted image of fruit by local sensing

2.10 Cucumber harvesting

> Research and development of a cucumber harvesting robot was conducted jointly by Okayama University, Ehime University and ISEKI & Co., Ltd. from 1991. Since there was no clear color difference between the cucumber fruit and the foliage of the cucumber plant, a two-wavelength type vision sensor using optical information in the visible and near infrared regions was developed.

In identifying fruit and foliage using a **vision sensor**, tomato, mandarin orange and grape can be identified using the RGB signals based on a color difference between the harvesting-stage fruit and the background. For example, a red-ripe tomato fruit has a high red reflectivity at 600–700 nm and a low green reflectivity at 530 nm. By contrast, a leaf has a low red reflectance and a high green reflectance. Accordingly, a fruit and a leaf can be differentiated by comparing the red color and the green color. It is more difficult to identify a fruit by RGB signal comparison in the case of cucumber, where the fruit and foliage have the same color.

However, **spectral reflectance**[I-2.2] measurements of various parts of a cucumber plant reveal two characteristic wavelength bands from the **visible region**[I-2.1] to the **near infrared region**;[I-2.1] the fruit has a lower reflectance at around 500 nm and a higher reflectance at around 750–900 nm than other parts of the plant.[24][I-2.2] Based on these characteristics, the **two-band vision sensor** of the cucumber harvesting robot for fruit recognition is comprised of a monochrome **CCD camera**[I-2.1] sensitive to both visible and near infrared waves and two **interference filters** of 550 nm and 850 nm.

Figure 1 shows the two-band vision sensor mounted on the cucumber harvesting robot.[II-3.14][25] Light is projected from either side of the object. The light reflected from the object passes through the interference filters to the monochrome CCD camera. Two optical filters[I-2.3] are used, one with a dominant wavelength of 548 nm, a maximum transmittance of 77 percent and a **bandwidth** of 12 nm and the other with a dominant wavelength of 851.6 nm, a maximum transmittance of 71 percent and a bandwidth of 4 nm. An image is acquired using each of these two interference filters (Figure 2). The image acquired using a 550 nm interference filter is called '550 nm image' and the image acquired using a 850 nm interference filter is called '850 nm image' hereafter. Computation using formula (1) for each pixel in the two images (Figure 3(a) and (b)) produces a **processed image** (Figure 3(c)). Since the fruit presents a high reflectance in the 850 nm image, its grayscale density value is high after processing, and the other parts present a lower reflectance in the 850 nm image, hence a low density value.

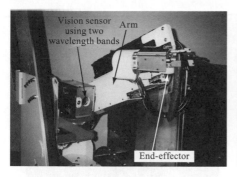

*Fig. 1 Two-wavelength-band
vision sensor*

*Fig. 2 Acquired image of
cucumber plant*

$$r = \frac{N_{850nm}}{N_{850nm} + N_{550nm}} \qquad (1)$$

N_{550nm} and N_{850nm} denote a density value for each pixel on the 550 nm and 850 nm images respectively. It is possible to extract cucumber fruit based on density differences. **Binarization**[1-2.4] on this basis can identify fruit and part of the front side of the leaf (Figure 3 (d)). However, the part of fruit with a greenish yellow tinge or specular reflectivity presents a low density value after processing and disappears when a **binary image**[1-2.4] is generated. To compensate for this, the 850 nm density image is used to add high density pixels to the binary image. This can more clearly delineate the shape of fruit.

Following noise removal, **filling**[1-2.4] and **labeling**, the feature quantities of each mass figure are extracted. The feature quantities include area, **Feret's diameter ratio**[1-2.4] and FLAG, where the presence of thirty or more consecutive lines, each having a horizontal width between 5 and 20 dots (inclusive), means FLAG = 1 and the absence of these lines mean FLAG = 0. Labels which satisfy the conditions of having an area of 300 or more and either a Feret's diameter ratio of 2.20 or more or FLAG = 1 are regarded as potential fruits and the center lines of these objects are extracted (Figure 4).

If a center line is 120 dots or more in length following noise removal and joining of neighboring lines, this pixel row is recognized as a fruit. If a pixel row is shorter, the system searches for other pixel rows within a range of 45 degrees below the first pixel row for joining. The first pixel row is joined with the most suitable pixel row among candidates and the process is repeated until the overall length reaches 120 dots or more. When there is no more pixel row below, it is recognized as a non-fruit (Movie). This system is capable of accurate extraction of feature quantities since it is less affected by the strength of light sources and the borderline of the fruit is clearly represented.

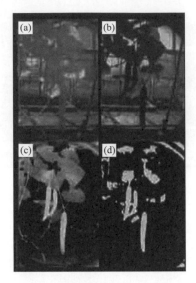

*Fig. 3 Images resulting from
fruit recognition algorithm*

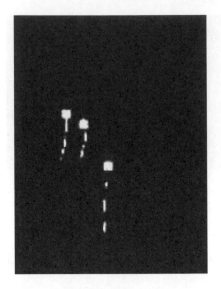

*Fig. 4 Extracted center lines of
cucumber fruits*

2.11 Cabbage harvesting

Research has pursued the mechanization of harvesting work for vegetables of large mass such as cabbage, as it puts a heavy burden on workers.[26] Some models for simultaneous harvesting of all crops have been put into practical use. A robot which selectively picked harvesting-stage cabbages was also developed. Position recognition and harvesting-stage determination using video cameras and 3-D vision sensors were tested.

The **cabbage harvesting robot**,[II-3.16] developed at NARC of the Ministry of Agriculture, Forestry and Fisheries, uses a color video camera to process an image of cabbages taken from above, estimates the 2-D positions and spherical diameters of their heads and identifies harvesting-stage cabbages based on the diameters.[27] The head of a cabbage has a similar color to the outer leaves and it is often partially covered by them. In image processing, RGB image data are converted into **hue**, **saturation** and **intensity** images[I-2.4] and the head is extracted by **binarization**[I-2.4] based on the **neural network** model. The images are matched with two cabbage model templates for estimation of the 2-D position and diameter of the head (Figure 1). The first template is used for identification of a focal area and the second template is used for a closer examination within this area. Eight CPUs perform high-speed parallel processing, which reportedly takes 8.8 seconds to process one image containing eleven cabbage plants and recognize their heads.

Research on ways to determine the correct stage for harvesting using a **3-D vision sensor**[I-2.5] has been conducted at Obihiro University of Agriculture and Veterinary Medicine.[28] This 3-D vision sensor consists of two laser range finders and a polygon mirror (a rotating hexagonal mirror) and is attached to the front of a tractor in the experiments. Scanning perpindicular to ridges is performed

 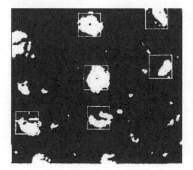

Fig. 1 Original image (left) and processed image (right) of cabbages

81

(a) 3-D shape measurement unit with laser range finder

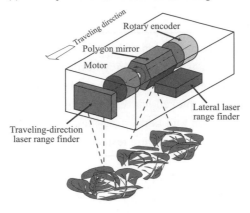

(b) Detection of diameter and partially unfolding leaves

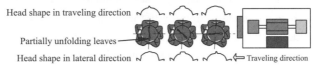

Fig. 2 3-D vision sensor

using a laser beam from a lateral **laser range finder** deflected by the rotating polygon mirror. Whole cabbages are scanned in this direction as the vehicle travels forward. A traveling-direction laser range finder scans the centers of cabbage heads parallel to ridges.

For determining the correct harvesting stage using 3-D vision sensors, three different approaches have been trialed: based on the spherical diameter of the

Fig. 3 Acquired image of cabbage
(variety: Early Ball)

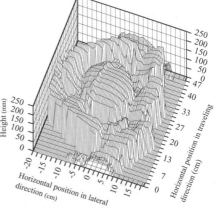

Fig. 4 3-D image of cabbage

head, the presence of partially unfolded leaves, and a combination of the first two approaches. The presence of partially unfolded leaves is used for the identification of harvest-ready crops since a cabbage nearing maturity tends to have partially unfolded leaves at the crown of its head as shown in Figure 3. Figure 4 is a **3-D image** of a cabbage acquired by polygon mirror scanning. Another approach being tested is a combination of the 3-D vision sensor image and the degree of hardness measured by compression of the cabbage head. Initial indications are that such an approach can achieve an accuracy of about 90 percent.

2.12 Orange grading

> Orange grading machines using image processing and optical sensing technologies have been in use at JA facilities in western Japan since around 1997. This section introduces the most popular model manufactured by SI Seiko Co., Ltd. which grades oranges based on images taken by TV cameras from six different angles. The latest technique used for the detection of rind puffing is X-ray image processing.

Citrus **fruit grading machines** have been mainly used for Unshu mandarin orange and late maturing orange varieties. The criteria for mandarin grading include shape, color, color irregularity, flaws, pest/disease damage, sunburn, rot and rind puffing. The criteria for grading late maturing varieties include granulation, caducous calyx (absence of calyx in *iyokan* etc.), pip number, rind thickness and many other items depending on the variety. The criteria that are detectable by imaging include color, color irregularity, flaws with a marked color difference from normal color, pest/disease damage and rind puffing. Unshu mandarins in particular present a wide range of colors from green in very early varieties to orange in late varieties, hence, the **TV cameras**[1-2.3] need to be carefully calibrated.

Figure 2 shows the flow chart of the image acquisition system[26] (Movie). A monochrome TV camera acquires an **X-ray image**[1-2.5] of the individual fruit projected onto a **scintillator**, then four laterally positioned color TV cameras take side view images of the fruit at 90-degree intervals. A camera installed above the conveyor line takes a top view image of the fruit, which is then rotated 180 degrees and for another camera to take a second top view image. Thus, images of all aspects of the fruit are acquired. The imaging computer (PC) has three capture boards, each of which can receive RGB signals from up to three cameras. After various types of imaging on the PC, feature quantities such as dimensions, color, shape, flaws and pest/disease damage are extracted and sent to the judgment computer controlling each conveyor line. The data processed on the imaging PCs are sent to the host PC, which stores the information.

Each TV camera has a **random trigger** mechanism by which it acquires an image at a **shutter speed**[1-2.3] of 1/1000th second at the moment of the fruit's passing across the **photo-interrupter**.[1-3.4] Since the conveyer speed is usually 60 m/min (1 m/s), a blur of up to 1 mm in images can result even at a shutter speed of 1/1000th second. While a faster shutter speed is desirable in order to minimize blurring, it would require a lighting system to provide greater illumination. The **resolution** of the camera is 0.35−0.4 mm and a **resolution** higher than **VGA**[1-2.1] is required to detect defects such as black spots that are smaller than 0.1 mm

Fig. 1 Various types of Unshu
mandarin oranges

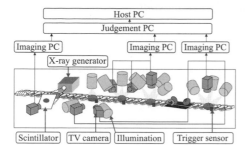

Fig. 2 Image acquisition system

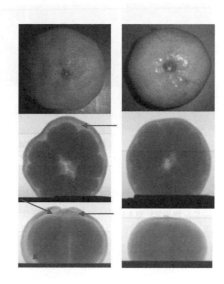

Fig. 3 X-ray images of puffy and
normal mandarin oranges

accurately. Figure 1 shows original images of fruits captured by the overhead camera. Each fruit is conveyed from the right end to the left end of each image and the image shows the moment a pulse (trigger signal) is generated by the photo-interrupter installed to the left side of the fruit just outside of the image.

Figure 3 shows the original and X-ray images of a normal mandarin and a puffy mandarin. Puffy parts of the mandarin rind are prone to rupture when packed fruit bumps against each other in cardboard boxes for shipment. Since fruit tends to rot from such damage, some producers exclude puffy fruits from high grade products. Severe puffing can be identified with the human eye, but it is difficult to detect on color images. The X-ray imaging system has been developed for this reason. Since X-ray beams are absorbed by water, puffy sections of the rind show a higher luminance than the dense fleshy part of the fruit. The arrows in Figure 3 indicate the sections of significant puffing. The current generation of citrus grading machines is fitted with a system which expresses the degree of rind puffiness numerically based on this density difference. The late maturing citrus fruits, as shown in Figure 4, can be graded on the same system as the different varieties of mandarin by simply changing the variety code.

Rotten mandarin fruit is removed manually in the first stage of grading on the current system but detection by imaging may be feasible. Since a fluorescent substance appears on oil glands in the rind of a rotten fruit, it can be detected by a black-and-white camera when illuminated by a light that excites wavelengths in

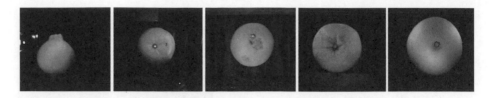

*Fig. 4 Acquired images of late maturing citrus fruits (from left, dekopon,
iyokan, kiyomi, hassaku and buntan)*

the ultraviolet region. Alternatively, a light can be located on the opposite side of
the camera to acquire a **transmittance image** as the light easily passes through
the collapsed oil glands of rotten fruit.[1-2.5]

2.13 Acid citrus grading

The grading system discussed in section 2.12 can be used for other citrus fruits such as *yuzu*, *sudachi* and *kabosu* with change of software. Internal quality criteria such as sugar content and rind puffing do not need to be checked for these citrus species – for their acids and juices as flavouring, rather than for their flesh – but external appearance is much more important than for the Unshu mandarin, for example, and very strict grading standards are applied.

Yuzu, *sudachi* and *kabosu* are acid citrus fruits from which the rind and juice are used in cooking, in contrast to the citrus fruits discussed in the previous section with pulp (juice vesicles) that is eaten. Accordingly, while the quality inspection items for acid citrus are similar to those for mandarin and late maturing oranges, a greater emphasis is given to external appearance rather than internal quality, with strict criteria for the apex (front) and peduncle (back) of the fruit. In particular, the apex of *yuzu* fruit is usually inspected more closely than the peduncle since its rind is used for cooking and garnishing and they are packed in boxes apex side up.

Yuzu are generally shipped a ripe-yellow but some are shipped as immature green fruit as in Figure 1(a) around August. They grow to their maximum size around September and begin to yellow by mid-October as shown in Figure 2(a), (b) and (c). Some fruits are immediately shipped after harvesting but many are stored at around 5°C for two to three months and shipped from November to January (especially around the winter solstice) after coloring in yellow. In this case, fruits are graded immediately after harvesting and stored according to the **grade and class** and often put through a **fruit grading machine** again prior to shipment. Attention is given to both the rind color and spotting as shown in Figure 2(c), as well as to discoloration of the calyx during storage.

 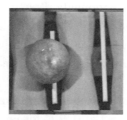 

(a) Green yuzu (peduncle end) (b) Sudachi (blossom end) (c) Kabosu (peduncle end)

Fig. 1 Acid citrus fruits: yuzu, sudachi and kabosu

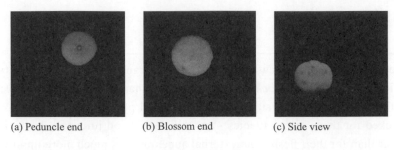

(a) Peduncle end   (b) Blossom end   (c) Side view

Fig. 2 Yellow color yuzu

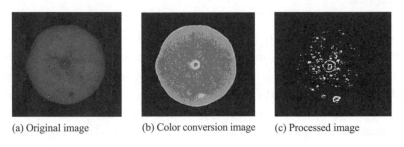

(a) Original image   (b) Color conversion image   (c) Processed image

Fig. 3 Citrus melanose

Since particularly strict external appearance standards are applied to *yuzu*, calyx recognition is used to identify the apex and the peduncle correctly. The fruits in four lateral view images are divided into two parts at the equator and the total size and number of defective areas are calculated for the apex side and the peduncle side for grading. In some cases, the front side criteria are used for the lateral view image of the apex side and the back side criteria are used for defective areas found in the lateral view images and the peduncle side images for grading. The **flower end** at the apex may be counted as a defect depending on the size.

Figure 3(a), (b) and (c) respectively show an original image of citrus melanose on *yuzu*, a color conversion image, and a binary image processed by inter-frame operation of the RGB color components, Prewitt operation and **binarization** with an appropriate **threshold value**.[1-2.4] Since black spots are larger on *yuzu* than Unshu mandarin and the camera's view angle can be reduced to enhance **resolution**, the system is likely to recognize them more accurately than in the case of the Unshu mandarin.

Sudachi and *kabosu* (Figure 1(b) and (c)) are available on the market most of the year since they can be stored at 0–2°C for up to four months but the peak shipment months are September and October. Young green fruits have a stronger flavor and a higher commercial value but yellowish fruits are less acidic and less

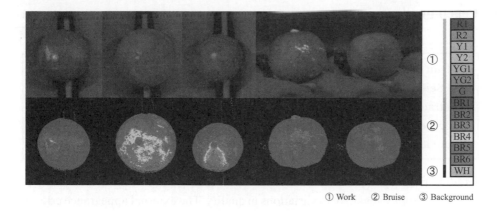

① Work   ② Bruise   ③ Background

Fig. 4 Defects of sudachi

valuable. Fruits of these species tend to have smaller size variations than the citrus fruits in the previous section. Since *sudachi* and *kabosu* are particularly small fruits, image **resolution** can be doubled when a lens with twice the focal length of the one used for standard citrus fruits is used (Figure 1 (b)). However, **TV cameras**[1-2.3] need to be set up carefully because many have a deep green color. When these green fruits are graded together with yellow *yuzu* fruits or mature fruits of other citrus varieties on the same system, the brightness setting of the TV cameras may have to be adjusted.

The top row images of Figure 4 are original images of various defects of *sudachi* fruits and the bottom row images are their color processed images. In color conversion, colors R1–G are assigned to normal areas, BR1–BR6 to defective areas and WH to the background. This example demonstrates that color variations in flaws are detectable by color conversion.

2.14 Peach grading (Movie)

A fruit grading robot with three-degrees-of-freedom arms and machine vision (SI Seiko Co., Ltd.) was commissioned at a JA facility in July 2002 (original contractor: Mitsui Mining & Smelting Co., Ltd.). It was the first such robot to succeed in fully measuring (six-view imaging) deciduous fruits and recording accurate information.

Grading peaches requires careful attention to quality as the commercial value of peaches is highly sensitive to variations in quality. The external appearance criteria for peach grading include dimensions, color, shape, flaws and pest/disease (codling moths, owlet moths, fruit cracking, bruising, scab, shot hole, twig rub, soiling, sunburn, chemical damage etc.). These problems are broadly divided into three groups: serious defects such as owlet moth and progressive soft rot, mild defects such as twig rub and soiling, and moderate defects including the rest. Different grading standards are applied depending on the extent of the defect.

Figure 1 is a system chart[30] of this **fruit grading robot**. Figure 2 shows six images (top, bottom and four side views) of a fruit (*Akatsuki cultivar*) captured by the **machine vision**[(I-2.1)] of this robot.[31(II-3.20)] The top left image was taken from above by a camera fixed above the conveyor which carries a fruit tray at a speed of 30 m/min. The bottom left image was taken from below by a swivel camera as the grading robot manipulated a fruit using a suction arm (⑤ in Figure 1). The remaining four images are side views that were taken by the swiveling camera which turns in the horizontal direction according to the robot arm movement and releases the shutter four times as the **rotational joint**[(I-3.6)] of the arm turns

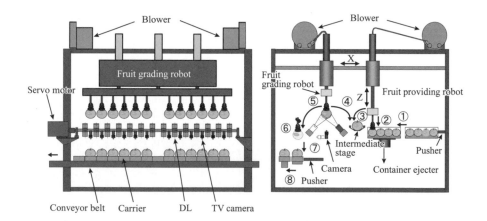

Fig. 1 Fruit grading robot system

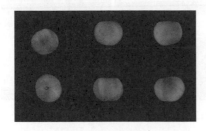

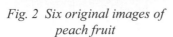

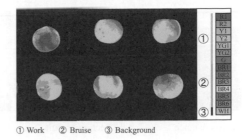

① Work   ② Bruise   ③ Background

Fig. 2 Six original images of
peach fruit

Fig. 3 Color conversion images

the fruit axially over a period of 0.6 seconds (⑥ in Figure 1). As shown in Figure 1, this robot can handle and image twelve fruits at once (Movie). The distance between the point at which the robot picks up a fruit with its **suction pad**[1-3.5] and the point at which it releases it is 1165 mm and the time it takes to perform this movement is approximately 2.7 seconds. The robot completes the whole process and returns to the default position in 4.25 seconds, including waiting time. This means that it can process three fruits per second, which is about ten times the amount of work performed by a person.

Figure 3 shows the images color-converted by **HSI (Hue-Saturation-Intensity) conversion**.[1-2.4] According to the color scheme shown on the right side of Figure 3, red, yellow and green colors indicate normal areas and blue colors indicate defective areas. Figure 4 shows defective areas (as white) based on an integrated assessment[1-2.4] which includes defects coded in blue by the HSI conversion as well as defects identified in other imaging processes. The frame surrounding each fruit in these images is the area of processing used in recognizing the fruit.

For internal quality, sugar content is measured by an **internal quality sensor** (Mitsui Mining and Smelting Co., Ltd.) using a near infrared light. Inspection for split pit is also desirable since it is difficult to recognize from external appearance. Technically, it is possible to use **X-ray images**[1-2.5] to detect this type of defect

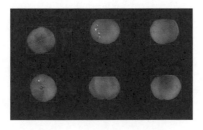

Fig. 4 Processed images

(a) Appearance   (b) Section of fruit   (c) Side view image   (d) Top view image

Fig. 5 Split pit of peach

as shown in Figure 5.[32] This technology is already in practical use in mandarin fruit grading.[II-2.12]

At the JA facility where this robot was first introduced, the number of workers employed in the fruit grading process has been reduced from about 140 to about 60 people. It is envisioned that this labor substitution can allow fruit growers who previously worked part-time at the **fruit grading facility** to devote their time to intelligent agricultural business, farm expansion or higher quality and safer cultivation management.

2.15 Pear grading

> The fruit grading robot described in 2.14 can be used for grading brown pear
> varieties such as *Kohsui*, *Hohsui* and *Nansui* and green pear varieties such
> as *Nijusseiki* by simply changing the variety code.

With a recent shift in consumer preferences from green pears such as *Nijusseiki* to
brown pears such as *Kohsui* and *Hohsui*, new value-added cultivars with a higher
sugar content such as *Nansui* are appearing on the market. Like peaches, pears are
graded on external appearance criteria, including dimensions, color, shape, flaws
and pest/disease damage. Defect items vary from variety to variety but they are
quite numerous: codling moth, owlet moth, physalospora canker, bruising, orange
peel, missing stem, calyx retention, mealy bug, bark miner, stink bug, water soaked
skin, black spot, scab, soiling, abrasion, frost damage, rust, chemical damage, fruit
cracking, leaf roller moth, lenticel and so on. Owlet moth and progressive soft rot
are classified as severe defects, abrasion, rust and soiling are classified as mild
defects, and various **grading criteria** are applied.[31] Figure 1 provides an example
of the grading criteria used at a JA facility on 24 September 2002. L_1 is the highest
grade and L_5 is the lowest grade.

In relation to color, images are processed by **HSI conversion**[1-2.4] and the whole
fruit and the top and bottom halves of fruit are graded based on five criteria,
including three **hue**[1-2.4] items, one coloration item and one **saturation**[1-2.4] item.
In brown pear varieties (especially Kohsui), subtle color variations need to be
quantified due to the presence of cork-colored lenticels all over the rind. In
relation to shape, three different criteria are used: ellipticity ($100 \times D_{max}/D_{min}$),

Color		L_1	L_2	L_3	L_4	L_5
Hue (H)	Whole	135–175	175	130–180	0–190	
Hue (H)	Top	135	135	130		
Hue (H)	Bottom	175	175	180	190	
Coloration (Red)	Bottom	58	63	68	80	
Saturation (S)	Whole	268	270	272	280	
Shape		L_1	L_2	L_3	L_4	L_5
Ellipticity		110	115	120		
Complexity		120	130	157		
Deformity		130	180	320		
Bruises		L_1	L_2	L_3	L_4	L_5
Severe	Top	4	20	50		
Severe	Bottom	6	30	70		
Moderate	Top	12	25	100		
Moderate	Bottom	15	30	130		
Slight	Top	15	40	110		
Slight	Bottom	30	60	160		

Fig. 1 Example of criteria for fruit appearance grading

complexity[1-2.4] (8 x P^2/A), and deformity (left-right symmetry of the bottom half of fruit), where D_{max}, D_{min}, P and A denote the maximum diameter, minimum diameter, perimeter and area of the fruit respectively. Flaws and pest/disease damage are recognized as such by the number of pixels in the areas that have abnormal rind colors after HSI conversion. If flaws and pest/disease damage have colors similar to the normal rind color, individual R, G and B images or their **processed images** are processed by a difference method and areas containing more than a set number of pixels at a density above a pre-determined threshold value are recognized as flaws and pest/disease damage. Detected defects are classified into serious defects, moderate defects and mild defects according to their color and shape. The top and bottom hemispheres are graded separately according to different criteria. Broadly speaking, serious defects in peaches, pears and apples include brown rot, physalospora canker, fresh bruise, fruit-piercing moths, codling moths and bird damage, moderate defects include compression/contusion bruising, bacterial shot hole, scab, Alternaria blotch, fly speck/sooty blotch, bitter pit, orange peel, sunburn, missing stem, stink bug, scale, fruit cracking and leaf roller moth, and mild defects include rust, twig rub and soiling. Any fruit which does not meet the criteria for a particular grade is downgraded. The items and reference values under these criteria vary depending on the fruit variety. The reference values concerning color and flaws (especially pest/disease damage) for the same variety also vary depending on the season in many cases. This means that the degree of fruit coloration and the type of pest/disease change depending on the time of shipment and the grading criteria must change accordingly.

Figure 2 shows the six images (top, bottom and four side images) of a fruit (Nansui variety) acquired by the **grading robot**.[II-3.20] For internal quality, sugar content is measured by an optic sensor (Mitsui Mining & Smelting Co., Ltd.) in

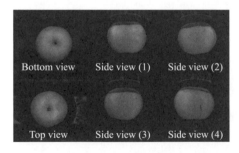

Fig. 2 Acquired images of pear fruit
(Nansui)

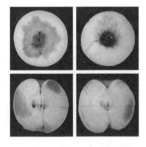

Fig. 3 Internally injured fruits of
Kohsui (upper images) and Hohsui
(lower images) (Mitsui Mining &
Smelting Co., Ltd.)

the same way as for peaches and apples. It is also possible to detect waterlogged or brown flesh as shown in Figure 3.

The table below shows the numbers of pears graded by each robot on 24 September 2002. Figure 4 is the result of the grading of Nansui fruits by lot on the same day. Take Lot Numbers 11 and 26 for example, No. 11 had more L_1 and L_2 fruits and fewer L_4 and L_5 fruits and No. 26 had the largest proportion of L_3, followed by L_2, L_4 and L_1 and the highest number of L_5 fruits of all lots on the day while both lots had similar numbers of fruit. The result shows large variations in the number of fruits and the grade ratio depending on the lot.

Table: Grading performance of robots (24 September 2002)

	Nijisseiki	Hohsui	Nansui	Total
Robot No. 1	8,282	0	17,768	2,605
Robot No. 2	10,050	0	16,513	2,656
Robot No. 3	0	8,263	17,514	2,577
Robot No. 4	0	6,930	16,485	2,341
Total	18,332	15,193	68,280	10,180

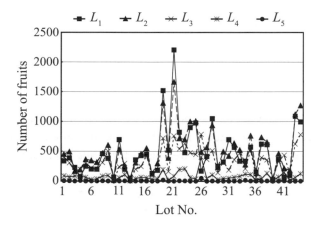

Fig. 4 Grading result by lot

2.16 Apple grading

> The grading robot described in the previous sections is capable of grading apples ranging in color from green to deep red (e.g., *Tsugaru, Ohrin, Yohkoh* and *Fuji*).

Each apple presents a mixture of a wide variety of colors from green to deep red. The season varies greatly for different varieties; for example, August to September for *Tsugaru*, October for *Ohrin* and *Yohkoh*, and November to December for *Fuji*. The appearance criteria are similar to peach and pear and major flaws and pest/ disease damage include rust, sunburn, Alternaria blotch, leaf roller moths, missing stems, bitter pits, stink bugs, scale, frost damage, fly specks, sooty blotch and ring rot. Like peaches and pears, progressive rot is classified as a severe defect and rust is classified as a mild defect.

The machine vision system[I-2.1]31 is mounted on the **fruit grading robot** and used under a **direct lighting system**[I-2.3] called DL (Figure 1, SI Seiko Co., Ltd.). It captures images using a color **TV camera**[I-2.3] with a **random trigger** function (effective sensor resolution 659 (H) x 494 (V), 1/2-inch **progressive scan single CCD**, RGB signal output) (Figure 2). With this robot, fruit is moved at a maximum speed of 1 m/s. Accordingly, the **shutter speed**[I-2.3] is set at 1/1000th second so that blurs in acquired images are limited to 1 mm. The RGB signals are transmitted from the camera to a computer via a capture board. The **resolution** of image data is 0.37 mm/pixel.

Figure 3 is the image processing system chart for this fruit grading robot.[II-3.20] Twelve swiveling cameras for the bottom and side view images capture images when they receive trigger signals from the robot controller. Three TV cameras connect with one PC via three capture boards for input and processing. Four PCs process a total of fifteen bottom and side view images (5 screens x 3 cameras) each during each stroke (4.25 seconds) of the fruit grading robot. PC-E processes three top view images per second while the fruit moves on the conveyor belt at 30 m/min after being released from the robot. The time required for image processing is 0.1 second or less per screen. In the end, a total of six view images are processed for each fruit and its dimensional, color, shape and flaw data are transferred to PC-J, which grades the fruit.

Figures 4 and 5 show images of an *Ohrin* apple and Figures 6 and 7 show images of a *Tsugaru* apple. For internal quality, sugar content is measured by a **spectrometer** (Mitsui Mining & Smelting Co., Ltd.) as in the cases of peaches and pears. The spectrometer also detects internal browning and water core. **X-ray imaging** or **X-ray CT** (Computerized Tomography)[I-2.5] have to be used for more

Fig. 1 DL

Fig. 2 Color TV camera

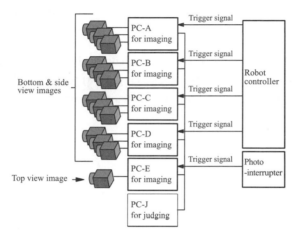

Fig. 3 Image acquisition system

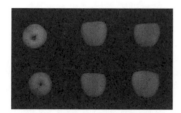

*Fig. 4 Acquired images of
Ohrin apple*

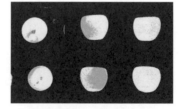

*Fig. 5 Color conversion
images of Ohrin apple*

*Fig. 6 Acquired images of
Tsugaru apple*

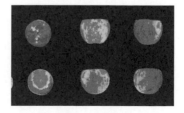

*Fig. 7 Color conversion
images of Tsugaru apple*

Fig. 8 Owlet moth damage to apple (Tsugaru)

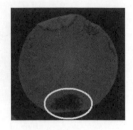

Fig. 9 Detection of owlet moth damage by X-ray CT

detailed internal quality data. The former can detect black heart and the latter can detect small spots in owlet moth damage (Figure 8) that are difficult to identify by examining the internal structure of the fruit (Figure 9).[32]

2.17 Persimmon grading

Some varieties of persimmon have a shape that is suitable for processing by the fruit grading robot described in the previous section but many JA facilities are using the orange grading system described in Section 2.12. Some producers are hoping to use machine vision for inspecting fresh and dried fruit in clear plastic packaging.

The shape of persimmon fruit varies greatly depending on variety, as does the level of astringency at the time of harvesting. Major sweet persimmon varieties include *Fuyu* (round), *Jiroh* (square) and *Nishimura-wase* (round). Astringent varieties include *Hiratanenashi* (square), *Tone* (square), *Saijoh* (conical) and *Atago* (conical). Different fruit grading criteria are adopted for different localities and cultivars but they generally include shape, color, color irregularity, flaws and pest/disease damage. The fruit grading system for persimmon is similar to that for citrus fruits (SI Seiko Co., Ltd.).

Figure 1 shows the apex and side views of a Tone fruit. The dimensions of the X mark at the center of the apex and the number and length of the black lines in the side view image are among the assessment criteria for grading. Figure 2 shows a **chromaticity conversion**[1-2.4] image and defect extraction image of the apex image on the left side of Figure 1.

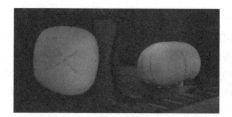 

Fig. 1 Apex view image and side view image of Tone fruit

Fig. 2 Color conversion image and defect extracted image of apex view

Figure 3 shows the peduncle view of the fruit and its chromaticity conversion image. The severe defect in the lower right corner of the fruit is expressed as a blue color. Since the calyx of persimmon is larger than that of other fruits, it is often included in the grading criteria. However, it is difficult to make correct judgment in image processing because the calyx can be partially withered as the one shown in the left side of Figure 4 which can be mistaken for a defect of the fruit itself shown in the right side of Figure 4.

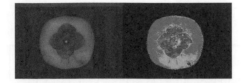

Fig. 3 Images of peduncle side

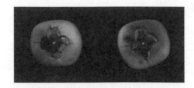

*Fig. 4 Withered calyx and
peduncle side defect*

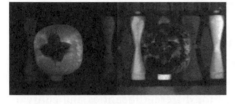

*Fig. 5 Fruit with bloom and its blue
component image*

*Fig. 6 Images of long-shaped
persimmon (Atago)*

Excessive bloom (Figure 5 left) and inky smudge (thin film of black powder) are sometimes added to the grading criteria. Bloom can be extracted clearly by blue component imaging in color imaging (Figure 5 right). Internal quality (sugar content) is detected by a **spectrometer** for citrus fruits based on the **near infrared spectrum method**.

Figure 6 shows the top and side views of an Atago fruit, which has a very different shape. These images were acquired by two top cameras and four side cameras as in the case of orange grading (Section 2.12). For fruits of this shape such as Atago, however, it is considered more appropriate to install two top cameras each before and after the fruit is turned over to take the side view images and two side cameras to capture images of the peduncle and apex sides.

It is becoming more common to vacuum pack persimmon fruits for a longer shelf life. Astringent varieties are packaged after the removal of astringency by alcohol or carbon dioxide. There is demand for automated grading of the color and defects of fruits packaged in this way. Figure 7 shows a packaged fruit. It requires a lighting system that does not cause **halation**. The image in Figure 7 was taken under a **direct lighting method** called DL (SI Seiko Co., Ltd.) and using a **PL filter** (polarizing filter) to remove specular reflection.[1-2.3] However, halation may not be removed completely due to re-**polarization** at creases in the packaging material.

Dried persimmon fruits have been used as preserved food for centuries. They can be graded by imaging since the fruit color changes according to the amount of

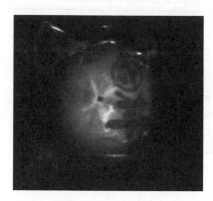

Fig. 7 Vacuum packed fruit   *Fig. 8 Images of dried long-shaped*
                                                    persimmon

bloom (Figure 8). Image-based grading itself is not difficult but the key question is how to measure all external aspects of a dried fruit in this case.

2.18 Tropical fruit grading

> The system used for grading citrus fruits is also used for grading tropical fruits such as wax apple and kiwi. The main grading criteria are sugar content and class.

The wax apple is a tropical fruit with a thick **cuticular layer**[1-2.2] as the name suggests. They are graded on the basis of appearance criteria such as color and shape as well as sugar content. As shown in Figure 1, the fruit has a bell shape which is approximately conical. The fruit grading system used in Taiwan has pin rollers like those used in the orange grading system (Section 2.12) but it only has one top view **TV camera**[1-2.3] and one side view camera and the fruit is not turned over.

Figure 2 shows color variations as the wax apple matures. The grade comes down from left to right. The green part at the apex is irrelevant to grade determination. The dark red color in Figure 2(a) falls into the highest grade but the use of an **indirect lighting method** such as diffuser panels or reflector panels allows the background to be included in the image and often makes it difficult to make subtle color measurement. A **direct lighting method**[1-2.3] with a better color reproducibility needs to be used in this case. The upper images in Figure 2 were acquired under dome lamps (indirect lighting) and the lower images were acquired under DL (direct lighting).

Figure 3 shows a color distribution chart of grade A, B and C wax apple fruits based on the HSI color conversion[1-2.4] images of the original images acquired using dome lighting and DL. The DL-based images resulted in a wide distribution of both the hue and saturation[1-2.4] values whereas the dome lighting-based images resulted in the overlapping of grade A fruits with grade B and C fruits. The **class** of agricultural produce is determined by the mass rather than the maximum

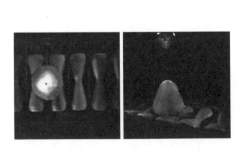

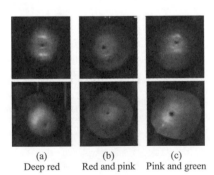

(a)   (b)   (c)
Deep red   Red and pink   Pink and green

Fig. 1 Top and side view images of wax apple

Fig. 2 Images acquired under dome lighting (upper) and DL (lower)

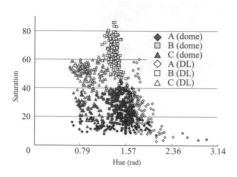

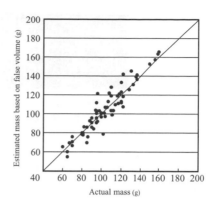

Fig. 3 Wax apple distribution after HSI
conversion

Fig. 4 Estimated mass

diameter or the projected area in many cases. The wax apple is one such case. Its class is determined by the following method. A conical model is constructed from the projected area and maximum diameter of a fruit acquired by two cameras for the calculation of a **false volume**.[I-2.4] The mass of the fruit is calculated by the false volume multiplied by an applicable specific gravity. The use of three or more cameras is desirable because the more cameras used, the more accurate the estimation becomes. However, some volume models can produce estimates with a coefficient of determination of R2 = 0.8820, a standard deviation of 9.2 g and a maximum error of 25.4 g using only two cameras as shown in Figure 4. One possible error is a cavity inside the wax apple as shown on the left side of Figure 5. Although still in the experimental stage, the use of a halogen lamp or other light source applied from below to obtain transmittance images is being trialed. Figure 6 demonstrates that light passes through hollow fruit better.

For kiwi fruit, a similar system with a sugar content sensor and one top view TV camera is in operation. Since the focus of kiwi fruit grading is on sugar content and class, other criteria such as shape, flaw, pest/disease damage and sunburn are not as important as for other fruits. Figure 7 shows the images of defective kiwi fruits. Most of the plant's body, including the fruits, have a high reflectance[I-2.2] in the **near infrared region**.[I-2.1] Consequently, the use of black-and-white TV

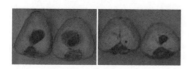

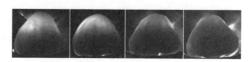

Fig. 5 Hollow fruits No. 1 & No. 2
and solid fruits No. 1 & No. 2

Fig. 6 Transmittance images (from left,
hollow No.1, No. 2, solid No. 1, No. 2)

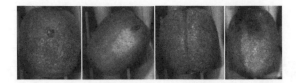

Fig. 7 Images of defective kiwi fruits

cameras with near infrared sensitivity may be sufficient for the grading of
fruits such as kiwi where subtle color variations are not very important in grade
determination. Figure 8 shows such an image. For sugar content, it is desirable
to be able to estimate the sugar content after force-ripening rather than at the
time of harvesting. Mango is another tropical fruit for which the introduction of
a non-destructive testing system is being considered.

Fig. 8 Image of kiwi fruit taken by
color-NIR TV camera

2.19 Grape cluster grading

Grapes are usually graded and packed at the farm but there is demand for automated grading of grape clusters in clear packaging so that information can be added to the products for a higher commercial value.

Mechanical grading is difficult for grapes since they grow in clusters. Packaging grapes is a particularly delicate work that requires attention to details. For this reason, growers have been grading and packaging grapes at the farm before delivering them to local JA facilities. However, there is demand for more objective grading of packaged grapes, as shown in Figure 1, at **cooperative fruit grading facilities** in order to increase uniformity in grading and add-value to the products. In this section, the grading criteria for *Kyoho* grapes are described.

The conditions of high grade *Kyoho* grapes include deep colors, bloom cover, large grape berries, no dry rachis, no rust, minimal exposure of inside rachis/peduncles and evenly formed cluster. Usually, a clear plastic film with the name of the grape variety and the local JA printed on it is attached to the package during grading at the farm. To make assessment through this film, a lighting system capable of preventing **halation**[1-2.3] is essential.

Figure 2 shows the images of grape clusters graded A, B and C with and without the plastic cover. To recognize such subtle color differences, it is advisable to use gamma correction in order to avoid non-linear color output.

Figure 3 shows only the **hue** values after **HSI conversion**[1-2.4] of the color images of ten random samples from each grade. They all have higher values without the plastic cover but their relative positions remain the same regardless of the presence or absence of the plastic cover. Figure 4 shows the **maximum length** and **breadth**[1-2.4] of each sample cluster, Figure 5 shows the bloom coverage (the ratio of the bloom-covered area to the total cluster area), and Figure 6 shows the rachis/peduncle exposure. All of them demonstrate that the feature quantities used as the grading criteria can be reproduced correctly in most cases regardless of the presence of the plastic cover except for the printing. For detection of bloom on *Kyoho* grapes, it is advisable to use a green component image that does not contain the fruit color since *Kyoho* grapes present with colors ranging from dark red to dark purple.

The box size for plastic packaged clusters also varies, most commonly between four and twenty clusters (1.5–8 kg) per box. Like grapes, strawberries are also graded and packaged in clear plastic packaging at farms but grading at cooperative fruit grading facilities has been trialed in recent years. The grading machine grasps individual strawberries by the peduncle to carry around in one example

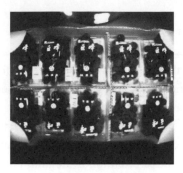

Fig. 1 Packed grapes

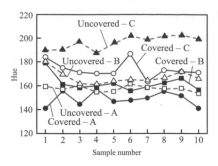

Fig. 3 HSI converted hue values

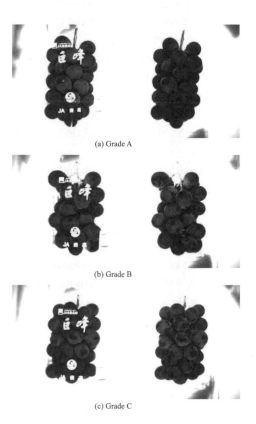

(a) Grade A

(b) Grade B

(c) Grade C

Fig. 2 Images of Kyoho grapes with and
without plastic cover

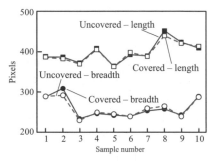

Fig. 4 Dimensions (Grade B)

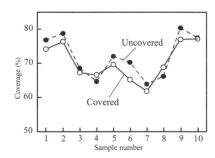

Fig. 5 Bloom coverage (Grade A)

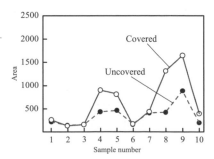

Fig. 6 Rachis/peduncle exposure
(Grade C)

(Tochigi Prefecture Agricultural Experiment Station & Nihon Kyodo Kikaku Co., Ltd.). Strawberries are carried on small tray conveyors in another example (YANMAR Co., Ltd.). Other fruits that are graded in boxes include cherries, plums, apricots and prunes.

2.20 Tomato grading

Two different automated systems are used for grading tomatoes: one uses roller pins such as those used for citrus fruit grading and the other employs a robot such as the one used for deciduous fruits. The tomato fruit grading robot was commissioned in 2004. Both systems assess each piece of fruit by acquiring six images.

The **grading criteria** for tomatoes include a wide variety of items such as color, color irregularity, discoloration ratio, shape, calyx splitting (calyx or fruit cracking), holes, stitching (zipper scar), **blossom end**, compression damage, streaking, meshing and soft fruit (puffiness). Although not all of these features can be identified accurately using imaging technology, the imaging processes for some of them are described below. The smoothness of the thick **cuticular layer**[1-2.2] of the tomato skin and the uneven shape of the fruit means that it is prone to cause specular reflection. In grading pre-cooled tomatoes, dew condensation on the skin often causes halation as well. The use of a direct lighting method[1-2.3] together with a PL filter[1-2.3] is recommended to avoid these problems.

There are three types of tomato grading systems, two of which are automated: the roller pin conveyor type used for grading citrus fruit, the three-degrees-of-freedom **Cartesian coordinate type**[1-3.6] robot used for grading peaches, pears and apples, and manual grading on tray conveyors. The first two types can extract information from six images of individual tomatoes. Their key requirement is maintenance of the correct postures of each piece of fruit. In manual grading, individual fruits are placed on the tray conveyor by hand and examined by one top camera or one top and one side camera only. The calyx side is graded by workers in a **tracking operation** (manual grading on the conveyor belt).

The most important of the six images is the one of the apex side. At some grading facilities, tomatoes are placed on the conveyor with the calyx side down and graded based on the image of the apex side only captured by the top camera. Figure 1 shows the sample images of the apex side. The average color values are derived from the values for each pixel obtained by chromaticity conversion and HSI conversion.[1-2.4] Color irregularity is calculated from the color value distribution of a given pixel. The extent of discoloration is calculated from the number (area) of pixels that have a red color below a certain value. Fruits with a blossom end that is larger than a certain size are downgraded. Shape-related feature quantities[1-2.4] such as **circularity factor, complexity, the ratio between max length and breadth**, and **the deformation from the gravity center** are derived from the apex side image. In particular, tomatoes with shapes such as

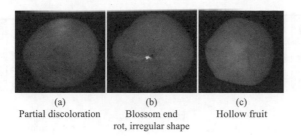

(a)   (b)   (c)
Partial discoloration   Blossom end   Hollow fruit
                             rot, irregular shape

Fig. 1 Images of apex side

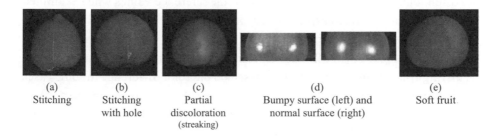

(a)   (b)   (c)   (d)   (e)
Stitching   Stitching   Partial   Bumpy surface (left) and   Soft fruit
             with hole   discoloration   normal surface (right)
                          (streaking)

Fig. 2 Side view images

those in Figure 1(b) and (c) tend to be hollow and are therefore downgraded. However, 2-D images need to be assessed carefully because even a slight tilting of the fruit can change the apparent shape of the fruit on the images.

Figure 2 shows some side images. The most notable defect on the side view is called stitching (Figure 2(a)). The length and number of stitches are among the grading criteria. Where stitching scars are so small (only one to two pixels on the TV camera[1-2.3]) that they cannot be extracted by color conversion, a first order difference (transverse **gradient**)[1-2.4] is used. Fruits that have stitching scars with holes are graded even lower. The depth of these holes cannot be measured by 2-D imaging but may be identified based on **intensity**[1-2.4] information. The fruit in Figure 2 (a) has a pointed apex, which is one of the grading criteria.

White streaking (Figure 2(c)) is said to be caused by potassium deficiency or bacterial wilt. A clear color difference is easily detectable but it is difficult to distinguish from color irregularity. One way to distinguish them is to check for fine bumps on the skin by touching with the fingers. Such bumps are difficult to detect under diffused light through the PL filter. Instead, it is necessary to acquire images under lighting conditions that produce specular reflection (Figure 2(d)). Soft fruit is a defect which exhibits subtle color variations (Figure 2(e)), and is very difficult to detect from external appearances. Internal defects of this type

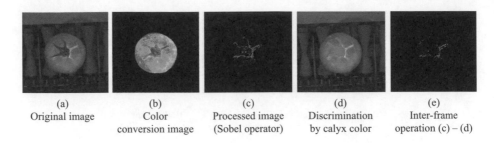

(a)	(b)	(c)	(d)	(e)
Original image	Color conversion image	Processed image (Sobel operator)	Discrimination by calyx color	Inter-frame operation (c) – (d)

Fig. 3 Image processing of calyx side images (skin cracking)

are often found to produce values outside of the reference values when examined by a near infrared **internal quality sensor**.

Figure 3(a) shows the original image of the apex side with skin cracks around the calyx and Figure 3(b) shows its chromaticity conversion. Skin cracking can be distinguished from the green calyx based on the color difference in some cases but it is not distinguishable solely based on the color difference in most cases. Accordingly, the image is processed using **edge detection filters** such as the **Sobel** operator to enhance the color difference.[1-2.4] This process results in extraction of the outline of the calyx as well as the skin cracks at the same time as shown in Figure 3(c). Figure 3(d) is an image of the green part of the calyx that is extracted and processed by **dilation**.[1-2.4] When Figure 3(d) is subtracted from Figure 3(c), the extracted image is almost solely of the defective part (Figure 3(e)). Figures 3(c), (d) and (e) are **binary images**[1-2.4] of the extracted green part.

A fruit grading system using rotary buckets for full measurement of long fruit such as the eggplant was developed by SI Seiko Co., Ltd. and introduced to JA's Okayama Binan facility in 2002. Six color and four black-and-white TV cameras are used on each conveyor line to identify dimensions, color, shape, flaws, pest/disease damage and dull fruit.

Grading long-shaped fruits (e.g., eggplant and cucumber) was previously done on the basis of imaging from above by a single **TV camera**[1-2.3] together with inspection of the bottom and side faces by the human eye. This approach relied on subjective judgments and was labor intensive. It could not provide fully **traceable** information as required to earn consumers' trust and confidence. To resolve these problems, a bucket tray with a mechanism to rotate eggplant and other long-shaped fruits 180 degrees (Rotary Basket, hereafter called 'RB') was developed and commissioned as part of an appearance grading system with color TV cameras which capture the images of fruits before and after they are rotated.[33, 34] In agricultural produce, the gloss or luster of the fruit skin is often considered to be an indicator of its quality. In the case of eggplant in particular, a lusterless fruit is called 'dull fruit' (an internal condition involving slow cell division caused by physiological disorder during growth) and loses its quality and commercial value significantly. Therefore, this system conducts **gloss measurement** by black-and-white TV cameras in addition to appearance assessment by color TV cameras.

Figure 1 shows the configuration of the fruit grading line. Fruits are fed into the roller type conveyor device, separated and put into RBs individually (cover plate A of the RB is down at this time as shown in Figure 2(a)). The top half of the fruit is filmed by three color TV cameras for external appearance and two black-and-white cameras for surface dullness. After this, cover plate A of the RB rises as in Figure 2(b) and closes as in (c). Lever B is released and the whole bucket rotates 180 degrees (as shown in (d) and (e)). Cover plate B is now positioned at the top. Lever A2 (A1) is released and cover plate B (A) opens as in (f). Lever C2 (C1) is released and cover plate B (A) descends as in (g). At this stage, the bottom side of the fruit is filmed. Once all the cameras (six color and four black-and-white) have finished imaging, cover plate B rises again. The bucket carries the fruit to an appropriate **grade and class** position, lever A1 (A2) is released, cover plate A (B) is opened (as shown in (h)), and the fruit is released onto a side conveyor. When cover plate A is on top, this fruit release action is the same as the cover plate opening action. Levers A1 and C1 open, close and lower cover plate A, and levers A2 and C2 open, close and lower cover plate B. They are operated point-symmetrically (Movie).

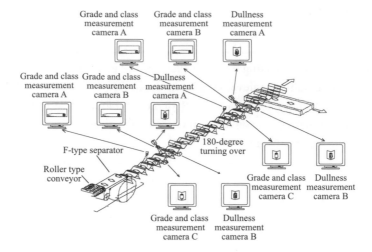

Fig. 1 Configuration of conveyors and cameras

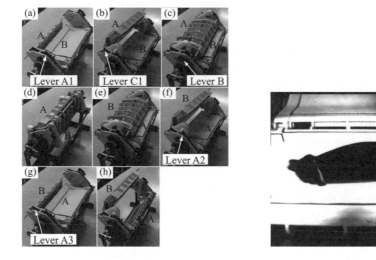

Fig. 2 Rotary bucket tray　　　　　*Fig. 3 Acquired image of eggplant*

Since eggplants have a gloss, like tomatoes, and a deep purple color, they easily reflect the background. The use of a **direct lighting method**[1-2.3] with a **PL filter**[1-2.3] can overcome this problem and achieve high color reproducibility in images (Figure 3). The conveyor travels at a speed of 30 m/min. Three cameras connected to one PC can process the images of three fruits per second (3 images per PC per second x 3 fruits = 9 images). The **class** factors that are detectable or measurable by the color TV cameras include maximum length, average diameter, area and **false volume**[1,2,4] of the fruit, and the grade factors include color, color irregularity, white calyx, green calyx, calyx area, C-curve, S-curve, crooked neck, cubic curve, maximum diameter, maximum-minimum diameter difference, apex dimensions ratio, irregular shape, total defect area, maximum defect area and average defect area. Injuries such as sunscald can be detected depending on the extent.

Figure 4 shows images of a glossy fruit and a dull fruit captured by the black-and-white camera. Since a three-**line light source** is used for lighting, it is possible to detect not only the degree of surface dullness but also a dull area of around 20 mm as shown on the right side of Figure 4 (more severe dullness near the calyx). This is detected before and after rotating the fruit by four monochrome cameras connected to one PC for image processing (4 images/second x 3 fruits = 12 images). The adaptation of this system for varieties of cucumber is anticipated.

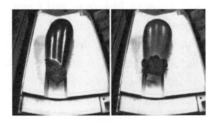

*Fig. 4 Glossy fruit (left) and dull
fruit (right)*

2.22 Cucumber grading

Shape sorting of cucumbers using black-and-white cameras has been in use for a long time. Images captured by the cameras from above are used in most cases and there is no inspection for flaws. Image-based grading for bitter cucumber is not as advanced as for other types of fruits but the key criterion for grading is maturity.

The **size class** of cucumber fruit is determined by its length and diameter while the **shape sorting**[35] into four or so grades is based on its shape characteristics such as curvature and thick blossom end (a difference between maximum and minimum diameters excluding 20 mm on both ends). The size of each conveyor tray is about 400 mm x 140 mm and in many cases is also used for eggplant and tomatoes. The conveyor travels at a speed of about 30 m/min and has a processing capacity of about 10,000 fruits per hour.

Figure 1 shows an example of the sizing criteria in relation to various shapes. They can be easily measured on images. Since color cameras have become as affordable as black-and-white cameras in recent years, items such as color, color irregularity, white defect and broken prickle are sometimes added to the grading criteria. There has been no report of image-based full measurement of cucumber for grading so far. The imaging and the imaging criteria have not changed very

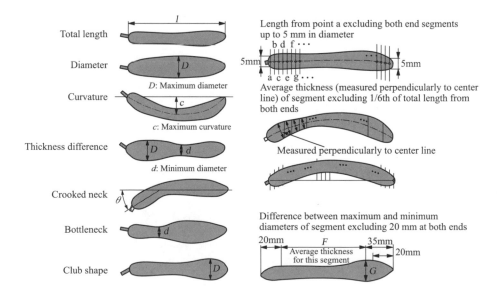

Fig. 1 Example of cucumber sizing criteria

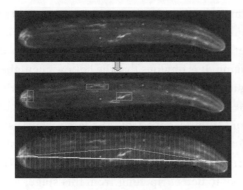

Fig. 2 *Acquired and processed images*
of cucumber

Fig. 3 *Cucumber boxing robot*

much over time, the main focus of the grading criteria is still the fruit shape and only the top side is inspected by imaging for color, flaws and pest/disease damage. Figure 2 shows examples of images. Once the **grade and class** of cucumbers are determined, they are arranged in trays and the **boxing robot** (Figure 3) grasps eighteen cucumbers simultaneously using **suction pads**[1-3.5] for packing.

Grading bitter cucumbers (Figure 4) has not been automated yet. At this point in time, in many facilities, the fruit is placed on a simple conveyor belt and visually sized and class-sorted using colored rulers attached to the line. Grading criteria include color, color irregularity, curvature and broken warts. However, hollowing of the fruit due to over ripening is considered to be a more serious problem than these external defects. The fruits must be immature when shipped, otherwise they ripen on their way to large markets such as Tokyo and their commercial value is reduced. Various over-ripeness sensors are being studied. One of them uses **transmittance images**.

Fig. 4 *Bitter cucumbers received by Japan Agriculture*

Figures 5–7 show images and transmittance images of bitter cucumbers at different stages of ripeness. The transmittance images are taken by an ordinary near infrared black-and-white camera by illuminating the cucumbers from the opposite side with a mirror halogen lamp. It is obvious that the hollow cucumbers have a high light transmittance (Figure 7). The difference is quite marked when dispersion is calculated because the warts of the bitter cucumber have a low transmittance. It is important to avoid light leakage from the lamp when using this method. Although fruits of markedly different ripeness can be distinguished by this method, even the cucumbers in Figure 5(a) and (b) already have yellowing seeds; thus it is desirable to find a way to detect ripening from this stage.

(a)   (b)   (c)

Fig. 5 Only slightly hollowed bitter cucumber fruits and transillumination image

(d)   (e)   (f)

Fig. 6 Moderately hollowed fruits and transmittance image

(g)   (h)   (i)

Fig. 7 Hollowed fruits and transmittance image

2.23 Sweet pepper grading

> The citrus fruit grading system is considered to be appropriate for grading sweet peppers at cooperative grading facilities. Tokyo University of Agricultural Technology has been conducting research on the new development of a mobile fruit grading robot since 2003 for the purpose of achieving information-intensive farm produce in precision agriculture.

Sweet peppers are generally graded on the conveyor type system used for citrus fruits. The **size class** is determined on the basis of mass, which is generally measured by **psuedo-volume**.[1-2.4] Grading criteria include color, color irregularity, shape and various pest/disease damage. Detection of thrips and oriental tobacco budworm larvae is particularly desired.

Like other fruits, measurement of all aspects of the fruit is important for sweet peppers. No system has been put to practical use as yet, but SI Seiko Co., Ltd. proposes the system depicted in Figure 1. First, randomly fed sweet peppers are arranged into a single line on the accumulation roller conveyor, of the type used for citrus fruit. The peppers are then arranged into the required orientation on an hourglass-shaped roller conveyor. They are then transferred to the pin roller conveyor where images are acquired by cameras as the peppers are rotated 180 degrees for full measurement. Figure 2 shows some of the images. Since oriental tobacco budworm larvae appear as simple dots on images taken by the **TV camera**,[1-2.3] other methods such as X-ray **transmittance imaging** are being considered but no definitive detection method has been found at this stage. For this reason, many producers are still trying to protect their crops by using yellow insect repellent lamps at night in the cultivation stage. The color image of green sweet pepper features a very small proportion of the blue component compared with other fruits (Figure 3), therefore many defects can be detected on a blue component image.

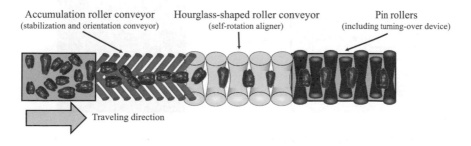

Accumulation roller conveyor
(stabilization and orientation conveyor)

Hourglass-shaped roller conveyor
(self-rotation aligner)

Pin rollers
(including turning-over device)

Traveling direction

Fig. 1 Fruit orientation conveyors

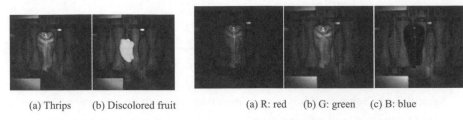

(a) Thrips   (b) Discolored fruit   (a) R: red   (b) G: green   (c) B: blue

Fig. 2 Side view images   *Fig. 3 RGB component images of Figure 2 (a)*

Tokyo University of Agricultural Technology is conducting research on the development of a **mobile fruit grading robot**.[36] The majority of conventional **fruit grading machines** have been stationary types installed at **cooperative fruit grading facilities**. The mobile robot will be able to provide information abut fruit product quality while simultaneously cross-linking it to farm information to prepare **yield maps** and **quality maps** of the farm.[37] It will also provide

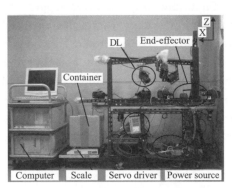

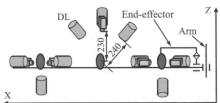

*Fig. 5 Configuration of TV cameras
and lighting devices*

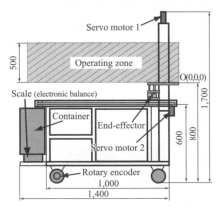

*Fig. 4 Mobile fruit grading robot
for sweet pepper*

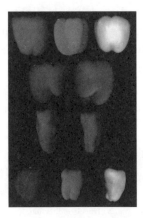

*Fig. 6 Images acquired by
mobile grading robot*

information about harvesting locations and times of individual fruit that can be used for tree management.[II-1.1]

This robot consists of a three-degrees-of-freedom **Cartesian coordinate arm**,[I-3.6] an end-effector, a **machine vision** system,[I-2.1] and a traveling mechanism. The machine vision system consists of four side cameras, one top camera and nine lamps (DL by SI Seiko Co., Ltd.).

Figure 6 shows examples of the images of four varieties of sweet peppers acquired by this robot. Figure 7 shows the relationship between the area of the fruit measured on the images captured by the five cameras and the mass measured by the electronic force balance. According to the diagram, all varieties except for the irregularly shaped Sonia have high coefficients of determination although there are minor variations between them.

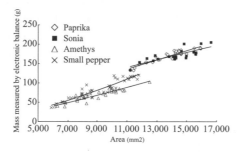

*Fig. 7 Relation between mass and area
          extracted from pepper images*

2.24 Potato and onion grading

> The fruit grading system for citrus is often used for grading potatoes but a smaller number of cameras (one to three) are used per conveyor line since size classification is more important in potato grading. Onions are difficult to measure by color TV cameras because the outer skin peels off during grading.

A wide variety of potatoes are available on the market in Japan, but they are mostly cultivars of *Danshaku* (Irish Cobbler) and *May Queen*. While the **size class** is the focus of potato grading, detection of grading criteria such as greening, flat shape, irregular shape, common scab, black dot and flaws is also desired. The same number of color **TV cameras**[1-2.3] as the citrus grading system (six per line) is required to measure these criteria accurately, but in many cases surface flaws and disease damage cannot be consistently detected due to the presence of dirt on the skin, peeling, surface knobs and dents.

Figures 1 and 2 are images of *Danshaku* and *May Queen* potatoes respectively. *Danshaku* cultivars tend to be rounder but they often have more blemishes on the surface than *May Queen* potatoes. For this reason, flaw and color detection is easier on *May Queen* potatoes. Since hollow potato detection is strongly desired for *Danshaku* potatoes, a line sensor system using a **soft X-ray**[1-2.5] (around 50 keV, 2 mA) (Izumi) has been introduced. Figure 3 shows examples of hollow potatoes and X-ray image.

The size class of potato is determined on the basis of mass, which is calculated from an accurate **psuedo-volume**[1-2.4] derived from images taken by multiple cameras. Broadly speaking, two cameras can estimate the mass with an error of less than 10 g or so (in the case of small potatoes). However, the mass density factor must be finely adjusted according to the variety and the false volume since potato sizes range from 40 g to 350 g.

Onions are very difficult for image processing. The shape and other features of an onion change dramatically as the outer skin peels off during machine processing. The following information is provided on the assumption that skin peeling does not occur. The grading criteria for onion include long shape, flat shape, multiple hearts, deformity, greening, sunburn, splitting, soiling, peeling and heart rot. To detect irregularly shaped onions, it is important to detect the orientation of the heart first. It is possible to determine the orientation by acquiring the outline of the onion and finding a line segment which connects a point of sharp variation and the center of area on the image. This makes it possible to determine the size class by the transverse diameter and to recognize

120

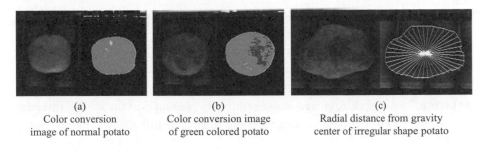

<table>
<tr><td align="center">(a)</td><td align="center">(b)</td><td align="center">(c)</td></tr>
<tr><td align="center">Color conversion
image of normal potato</td><td align="center">Color conversion image
of green colored potato</td><td align="center">Radial distance from gravity
center of irregular shape potato</td></tr>
</table>

Fig. 1 Sample images of Danshaku potatoes (SI Seiko Co., Ltd.)

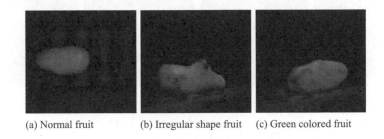

(a) Normal fruit   (b) Irregular shape fruit   (c) Green colored fruit

Fig. 2 Sample images of May Queen potatoes

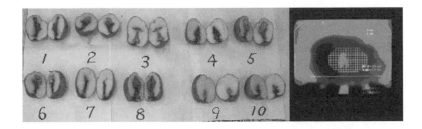

Fig. 3 Hollow potatoes and X-ray image (Tekko Co., Ltd.)

the long onion (long diameter in the direction of the heart) and the short onion (short diameter in the direction of the heart).

Figure 4 shows acquired images of onions. These images mostly provide skin information and onions often exhibit very different features when even one layer of outer skin is peeled off. Unlike potatoes, onions tend to produce severe **halation**,[1-2.3] which requires some reduction measures. One of the internal quality defects of the onion is heart rot and the use of soft X-ray is being trialed to detect it.

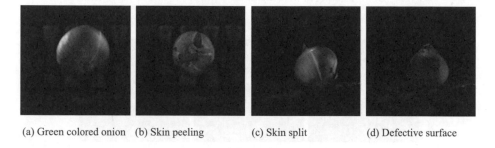

(a) Green colored onion   (b) Skin peeling   (c) Skin split   (d) Defective surface

Fig. 4 Sample images of onions

2.25 Asparagus grading

> Asparagus grading machines with special trays and conveyors have been commercialized by several manufacturers. For asparagus, images of the whole asparagus, enlarged images of the spear and images of the entire circumference are important.

Asparagus can be graded using images. The grading criteria include color, overall curvature, spear tip curling, open tip, spear ratio, purple stem end, white stem end, flat stalk, wilted bract, split stalk and broken stalk. Although the **size class** can be determined on the basis of the mass obtained from a **psuedo-volume** in imaging, more accurate measurement can be obtained by fitting a load cell to the conveyor tray. Several manufacturers are marketing various types of grading machines with different capacities but the most common models have a conveyor speed of around 30 m/min and a processing capacity of six asparagus per second. Many of the systems have two to three **TV cameras**[1-2.3] per conveyor line.

Grading work involves manually feeding the asparagus from containers into the grading machine and cutting them to an overall length of approximately 25 cm. Once the grade and class are determined by image processing, the products are sorted. Several spears of asparagus are bundled together to make a bunch of approximately 100 g. To maintain a good presentation, the tips are bound with a piece of tape and the stems are bound with a rubber band. This work stage is robotized and the order of binding varies depending on the manufacturer. Higher grade products are generally classified into S (11–14 spears), M (7–10 spears), L (4–6 spears), 2L (3 spears) and 3L (2 spears). Lower grade products have fewer size classes. Bundles are shipped in boxes such as a 5 kg Styrofoam box.

Figure 1 shows an example of the asparagus image processing system. Three **line light sources** fitted with **PL filters**[1-2.3] and two color TV cameras are used to take images of the entire asparagus and the spear tip. Although only two cameras are used in this example, a camera to capture an image of the entire asparagus from the direction of the spear tip camera and a camera to capture an axial view image from the stem end can be added to this setup. The additional cameras are expected to improve detection of defects, including detection of flat and split stems.

Figure 2 shows images of asparagus acquired by the two cameras and the processed images. The difference between the center line of the stalk and the line connecting the two ends of the stalk is measured to identify curvature and either the **complexity**[1-2.4] or the variances of the image intensity values are used for open spear tips. The images in this example were acquired by cameras having

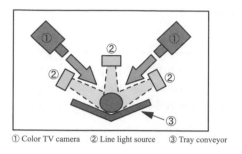

① Color TV camera   ② Line light source   ③ Tray conveyor

*Fig. 1 Image processing system for
asparagus (SI Seiko Co., Ltd.)*

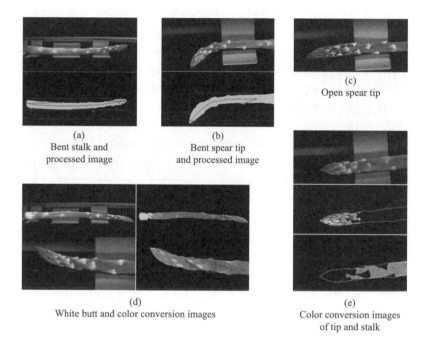

(c)
Open spear tip

(a)
Bent stalk and
processed image

(b)
Bent spear tip
and processed image

(d)
White butt and color conversion images

(e)
Color conversion images
of tip and stalk

Fig. 2 Acquired images and processed images

VGA[1-2.1]-class **resolution** but the use of **XGA**[1-2.1]-class cameras can eliminate the
need for zooming in the spear tip image and only two cameras may be required
to acquire all images except for the axial view of the stem.

The question here is whether it is possible to take measurements of the entire
circumference of the vegetable which is also important for other types of fruits.
There has been no commercial model which has satisfied the requirement to
turn over the asparagus and take accurate measurement of the bottom side so
far. However, Figure 3 depicts one possible approach for measuring the top and

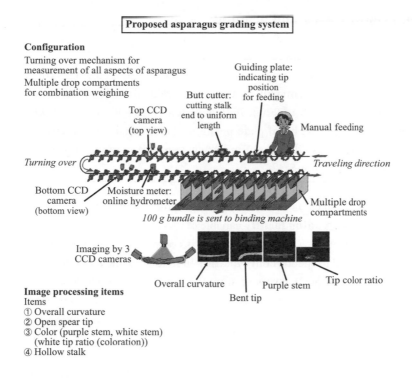

| Proposed asparagus grading system |

Configuration

Turning over mechanism for measurement of all aspects of asparagus

Multiple drop compartments for combination weighing

Guiding plate: indicating tip position for feeding

Butt cutter: cutting stalk end to uniform length

Top CCD camera (top view)

Manual feeding

Turning over

Traveling direction

Bottom CCD camera (bottom view)

Moisture meter: online hydrometer

100 g bundle is sent to binding machine

Multiple drop compartments

Imaging by 3 CCD cameras

Overall curvature

Bent tip

Purple stem

Tip color ratio

Image processing items
Items
① Overall curvature
② Open spear tip
③ Color (purple stem, white stem) (white tip ratio (coloration))
④ Hollow stalk

Fig. 3 Asparagus grading system (SI Seiko Co., Ltd.)

bottom sides under the same condition. Using this mechanism, the top side of the asparagus is measured while it is on the conveyor tray and the bottom side is measured while it is suspended. This mechanism has already been put to practical use as part of a leek grading machine.[II-2.26]

For leeks, post-harvest preprocessing work such as root cutting, skin peeling and grading (based primarily on white stem-related criteria) were automated by SI Seiko Co., Ltd. in 2002. A total of ten color and black-and-white TV cameras measure the basal plate position, the dimensions of the white stem and the pest/disease damage. Similar systems are coming into wide use in leek producing regions.

Leeks (long white leeks) require extensive preprocessing work after harvesting. A **cooperative fruit grading facility** capable of this preprocessing work prior to grading began operation in 2002.[38] This facility is comprised of the preprocessing stage where leeks are fed to special conveyor trays individually and automatically trimmed of roots, leaves and outer layers of skin, the grading stage where leeks are sorted by **grade and class** by the image processing system, the binding and sealing stage where leaks are bound and taped into a bundle of two to eleven plants and packed in a box of ten bundles, the shipping stage where boxed leeks are loaded onto pallets by robots, and the carbonization process where waste is recycled into **soil conditioner**. The automated preprocessing unit has a capacity of 10,000 leeks/h and can cut the number of preprocessing workers by 30–40 percent as well as the level of production labor to around 40 percent of that required by non-automated processes.

Figure 1 is a schematic diagram of preprocessing procedures. First, lower and dry leaves are removed from dirty leeks as shown in Figure 2. The stem cutting position is identified by the **TV cameras**[1-2.3] and the root section is cut off accurately. Figure 3 shows basal plates that have been cut correctly. If too much of the basal plate is cut off, the heart of the leak is exposed, and if too little,

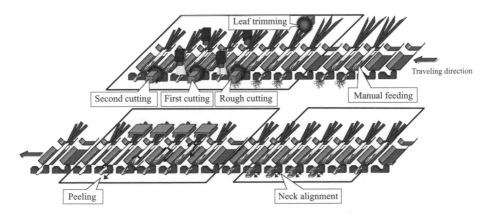

Fig. 1 Preprocessing procedures for leek grading (SI Seiko Co., Ltd.)

Fig. 2 Received leeks

Fig. 3 Basal plate

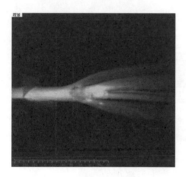

Fig. 4 Image of leek neck

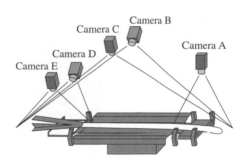

Fig. 5 Camera positions for top side images

peeling becomes difficult. Figure 4 shows a white dotted line indicating the leaf base (neck) position detected by a color TV camera at the exit of the root cutting unit. Outer layers of the skin can be peeled off by application of pressurized air from a blower moving from the neck position to the basal plate. Once all leeks are aligned according to the detected neck position, this peeling unit can process twenty-one leeks simultaneously. This system has many unique mechanisms such as an oscillating grip to handle bent, short and thin leeks. After the preprocessing, leeks move on to the grading stage (Movie 1).

Figure 5 shows the grading camera positions for the top views. Camera A (black-and-white) detects stem diameter, cameras B and C (color) measure white stalk length and curvature, and cameras D and E (black-and-white) detect leaf number, leaf defects and pest/disease damage. Figure 6 shows images taken by cameras A, B and D. The conveyor trays turn over, and then two black-and-white cameras (F and G) take images of the bottom side of the leek. Figure 7 shows an overall view of the preprocessing and grading system.

A preprocessing machine for personal use, which was developed by the Urgent Project 21, has also been put to practical use (Movie 2).

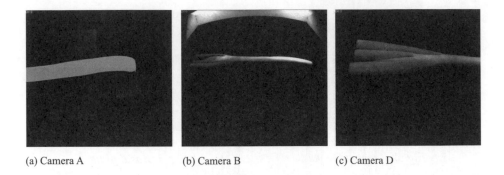

(a) Camera A   (b) Camera B   (c) Camera D

Fig. 6 Images at grading stage

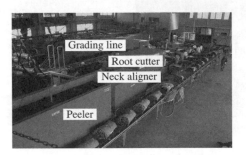

Fig. 7 Leek processing and grading system

2.27 Cabbage preprocessing and large fruit grading (Movie)

> Machines fitted with large trays are often used for handling large, almost spherical fruits such as cabbages and water melons. A cabbage preprocessing machine which identifies the color and cuts off the outer leaves was developed by BRAIN and YANMAR Co., Ltd. as part of the Urgent Project 21 in 2002. Sample images of melons and pineapples are also shown in this section.

Harvested cabbages have many outer leaves, as shown in Figure 1(a) and (b), which are removed, as shown in Figure 1(c), prior to shipment in order to reduce their apparent volume. A machine to complete this preprocessing work was developed by the Urgent Project 21 in 2002.[39] This system, shown in Figure 2, feeds cabbages one by one to the tray conveyor in Figure 3, identifies the color of the bottom part of the cabbage using a color **TV camera**[1-2.3] above, and cuts outer leaves with the rotary blade of the second cutter based on imaging data. There are three levels to the rotary blade action (deep cut, shallow cut and no cut) according to the color of the outer leaves. At a conveyor speed of 30 m/min, the system can process up to one cabbage per second. For other leafy vegetables, such as lettuce, some facilities are using imaging-based grading machines which easily measure the **grade and size**.

While internal quality criteria such as sugar content and hollowing are emphasized in the grading of melons, their color and shape must be checked by imaging as well. In particular, some muskmelon producing regions are hoping to grade their products based on the netting pattern on the rind. A high grade muskmelon is almost completely globe-shaped and covered with elaborately reticulated fine ribs of uniform height (Figures 4 & 5). Studies on net pattern recognition by way of analysis of **textural features**[1-2.4] such as ASM, CON and

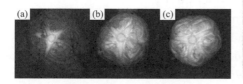

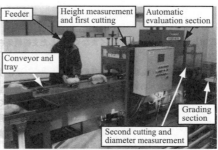

Fig. 1 Images of cabbage
(a) Requiring deep cutting
(b) Requiring shallow cutting
(c) Requiring no cutting

Fig. 2 Cabbage trimming machine[40]
(BRAIN, YYANMAR Co., Ltd.)

129

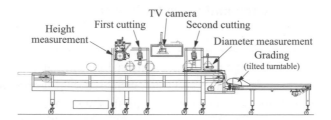

Fig. 3 Outline of cabbage trimming machine (40) (BRAIN, YANMAR Co., Ltd.)

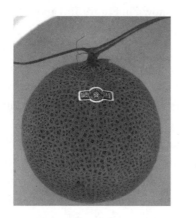

(a) High grade   (b) Low grade

Fig. 4 Ideal net on melon rind   *Fig. 5 Binary images of melon net (4 cm x 4 cm) (Tokyo University of Agriculture)*

COR and **fractal** analysis have been reported.[40, 41] Some melon **grading machines** use the conveyor trays and image processing systems such as those used for tomatoes and other fruit (Movie).

A study of sugar distribution in melons attempted to visually represent the sugar level of each pixel by capturing cross-section images of a halved melon in the 800–1100 nm range by every 5 nm using a **multi-spectral camera**.[(I-2.5)42] The image size was 384 x 192 pixels and the time of exposure at each wavelength was varied between 0.2–5 seconds according to sensitivity.

An **MRI**[(I-2.5)] image-based detection method for hollowing and sugar content of water melon has been reported,[43] in addition to other methods of detection by percussion, electrical impedance[44] and density.[(I-3.2)45] For boxing water melons, a **Cartesian coordinate type**[(I-3.6)] robot with an end-effector fitted with a suction pad[(I-3.5)] is already in operation.

With pineapples, it is often difficult to determine maturity and internal disease damage from external appearance alone (Figure 6). Therefore, an internal quality inspection system using **transmittance imaging** as shown in Figure 7 has been

Fig. 6 Pineapples with various maturity and internal quality (BRAIN)
Left: Internal disease.   Top right: Overmature fruits.   Bottom right: Immature fruits.

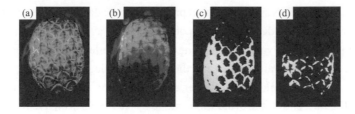

Fig. 7 Transmittance images and binary images (BRAIN)
(a) Normal fruit.   (b) Fruit with internal disease.   (c) Optimum maturity fruit.   (d) Immature fruit.

developed.[46, 47] This is based on the facts that the amount of transmitted light increases with maturity and that the light transmittance varies when the fruit is internally damaged. Research on a portable sensor is underway so that this system can be used outdoors.[48, 49]

2.28 Chrysanthemum cut flower grading　　　　(Movie)

> An automated grading and packing facility for cut standard chrysanthemums was set up by Panasonic Corporation. and Nishijima Co., Ltd. in Atsumi, Aichi Prefecture, and has been in operation since 1997. Work procedures such as receiving, image-based grading and packing have been robotized. Like standard chrysanthemums, the possibility of image-based evaluation is being investigated for spray chrysanthemums (spray-mums).

The quality of a cut flower is determined by many factors such as foliage balance, neck length, stem curvature, and color and luster of flower and foliage as well as size and mass. However, these standards are ambiguous and the flowers are often evaluated in bunches at auctions. The price is unstable since it is set by individual valuers (bidders) and each bidder often sets subtly different prices from one day to another. Cut standard chrysanthemums are graded on the basis of criteria such as overall height, main stem diameter, main stem curvature, neck length, inter-node length, leaf area, leaf wilt, balance between foliage and flower and the colors of foliage and flower.

Figure 1 shows examples of cut flower appearance and scores given by an expert out of 100. Figure 2 shows the evaluations made by two experts on five representative samples from nine different cultivation and treatment plots. It demonstrates that the evaluation scores of the two experts do not necessarily follow the same pattern and that each expert has given different scores to the same flowers in a second round of evaluation in some cases.[50]

For this reason, the use of **machine vision**[1-2.1] for more objective evaluation is being studied.[51] The system can extract feature quantities such as overall height, leaf area, main stem diameter, main stem curvature, neck length and first leaf length from images and evaluate them using a **neural network**. Figure 3 shows images after **binarization**[1-2.4] of some of the samples from those plots. The numbers in brackets are the scores given by Expert 1.

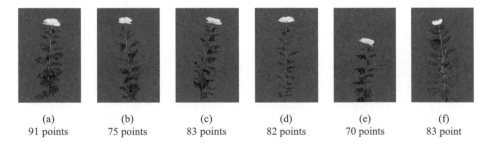

| (a) | (b) | (c) | (d) | (e) | (f) |
| 91 points | 75 points | 83 points | 82 points | 70 points | 83 point |

Fig. 1 Chrysanthemum cut flower appearance and scores

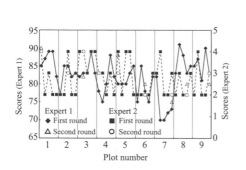

Fig. 2 Evaluation scores given by two
experts to 5 samples from each plot

Fig. 3 Binary image sample from each
plot (score in brackets)

The volume of chrysanthemum production is the highest among all cut flowers and the value of chrysanthemum production accounts for about 35 percent of the total cut flower production in Japan. A specialized **cooperative flower grading facility** (Mum-port) in the town of Atsumi, Aichi Prefecture, has been in operation for several years. Chrysanthemum flowers (150 stems per unit) from each producer are received at the facility using an ID card and carried to feeder conveyors and grading conveyors by robots (Figure 4). They are cut to a length of 87 cm and have their lower leaves removed. Then, **TV cameras**[1-2.3] capture their images for color, shape and foliage balance and sort them into eight **grades and classes** plus the off-grades. The system takes 0.6 seconds to process one stem. After stem cutting in water and conditioning for water uptake, the cut flowers are packed in a box of 200 stems by transfer robots (Figure 5) (Movie).

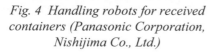

Fig. 4 Handling robots for received
containers (Panasonic Corporation,
Nishijima Co., Ltd.)

Fig. 5 Inside view of Mum-port
(Panasonic Corporation,
Ltd., Nishijima Co., Ltd.)

Processing up to 1900 boxes per day, this cooperative flower grading facility handles 99 million stems per year. The boxes of chrysanthemums are labeled, weighed, packed automatically and carried to cold storage.

The evaluation of spray type flowers or inflorescences such as *Gypsophila paniculata* and *Statice limonium* is slightly different from that of standard chrysanthemum in that the multiple flower formation and the number of florets must be included in the criteria. However, research on their image-based evaluation by a similar approach is underway and a **quality evaluation algorithm** (Figure 7) for image processing of spray chrysanthemums (Figure 6) has been reported.[52–54] The original image is binarized at two different levels to generate one **binary image**[(1-2.4)] of the entire stem and one binary image of the flower only. Then, the system detects the lowest node of the stem O and removes the main stem from the entire stem image. It finds the approximated polygon of the stem image and the flower image respectively and evaluates spray formation and foliage balance.

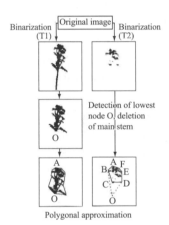

Fig. 6 *Spray chrysanthemum*   Fig. 7 *Image processing algorithm*
                                                            for spray-chrysanthemum
                                                            (Okayama University)

2.29 Wood Inspection

A high-speed image processing system for determining timber orientation and detecting laminar wood defects is in practical use. The defect inspection criteria include dead knots, missing knots, live knots, partially missing knots, splits, rot, want, wane and gum deposits.

Timber production generally entails the following procedures. The raw wood is brought to the yard for debarking. Careful inspection of the external condition of the timber is carried out, the most efficient conversion method is determined, and the timber is sawn into square pieces. After drying and curing, the timber pieces are planed and finished to designated dimensions. Only those which meet the standard **moisture content** and **Young modulus**[1-3,2] are shipped.

First, the orientation (inside and outside) of the sawn wood (Figure 1) needs to be determined based on annual rings prior to drying. The side of the older rings is the outside and the wood pieces are stacked with the outside up. There are several reasons for this. Since wood tends to warp after drying, as shown in Figure 2, wood stacked in this manner is more stable when being planed. Stacking all pieces in this manner after drying can minimize warping. And it is convenient to know which side is the inside or the outside when producing laminar wood sheets. The inspection system is shown in Figure 3. It has a conveyor speed of 60–90 m/min and can process up to three pieces per second. However, it requires careful monitoring since it is difficult to determine the orientation of some pieces, as shown in Figure 4. Figure 5 shows an example of wood laminate glued together in the correct orientation.

Figure 6 shows an example of image processing system configuration for laminar wood, planed laminated wood and solid natural wood. It has six black-and-white **TV cameras**.[1-2,3] The middle cameras inspect defects such as dead knots, missing knots, live knots and cracks. The cameras above and below look for partially missing knots, want and wane. The system handles wood pieces from

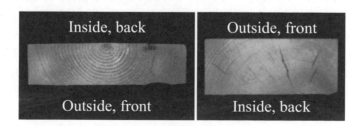

Fig. 1 Samples of sawn wood

135

Fig. 2 Warped sawn wood after drying   *Fig. 3 Growth ring inspection system
(SI Seiko Co., Ltd.)*

*Fig. 4 Sawn wood of unclear
orientation*

Fig. 5 Laminar wood

2.5 m to 6 m in length and has a conveyor speed of 120 m/min. Since the 1/2-inch
CCD cameras[1-2.1] are fitted with a 6-mm lens and set at a distance of about 23
cm to the target object, it has a visual field of over 20 cm. Accordingly, thirty
trigger sensors are installed at intervals of 20 cm in order to inspect the entire
length of a 6-meter piece. The **shutter speed**[1-2.3] is set at $1/1000^{th}$ second for work
moving at 2 meters per second. Figure 7 shows a wood piece at the entrance to
the inspection system and Figure 8 shows examples of missing knots, partially
missing knots and split defects and their images. Laminar wood may exhibit an
uneven surface (Figure 9) depending on the accuracy of lamination. This defect
is inspected using a **laser displacement meter** installed at a later stage on the
inspection line.[1-2.5]

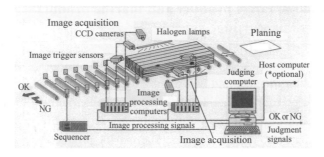

Fig. 6 Image inspection system (SI Seiko Co., Ltd.)

Fig. 7 Entrance of inspection system (SI Seiko Co., Ltd.)

Fig. 9 Uneven surface of laminated wood

(a) Missing knot

(b) Partially missing knot

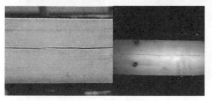

(c) Split

Fig. 8 Various defects and their images

2.30 Animal product, forest product and marine product inspection and processing

> The use of color imaging and X-ray imaging has been trialed for inspecting and evaluating meat products. Image data often makes automation of various processing procedures for marine products easier. Inspection and processing approaches for various animal, forest and marine products are described in this section.

Beef quality is determined on the basis of marbling pattern, aging, and red meat-fat-bone ratio. A study on a technique to extract red meat from images, as shown in Figure 1, uses color and **texture** to detect the extent of marbling.[55, 56] Color image processing[57] is also used to detect defects for chicken meat in automated inspection systems which have been in use in the U.S. for many years. In Figure 2, abnormal parts are indicated by different density values. Defect detection using **hyperspectral imaging** has been reported recently.[58] At the same time, recognition of different cuts of chicken such as drumsticks, keels, ribs, thighs and wings is also being trialed. Figure 3 shows images of various cuts acquired by color TV cameras.[1-2.3] While it is sometimes difficult to distinguish between keels, ribs and thighs on color images due to a wide range of shapes and colors, they are often distinguishable more readily on **X-ray images**,[1-2.5] which reveal the bones inside the meat.

In the egg industry, there is a high demand for technology to detect blood in eggs due to consumer preferences. Sample images in Figure 4 were captured by five color TV cameras with the eggs illuminated from below by halogen lamps. Figure 4 shows both normal and abnormal eggs. **Transmittance images** are shown at the top and **chromaticity conversion**[1-2.4] images at the bottom. The four images on the left were taken by side view cameras at 90 degree intervals and the image at the far right was taken by a top view camera from above. While it is

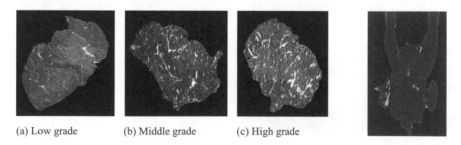

(a) Low grade   (b) Middle grade   (c) High grade

Fig. 1 Processed images of beef (Sung Kyun Kwan University, Korea)

Fig. 2 Chicken defect detection[58]

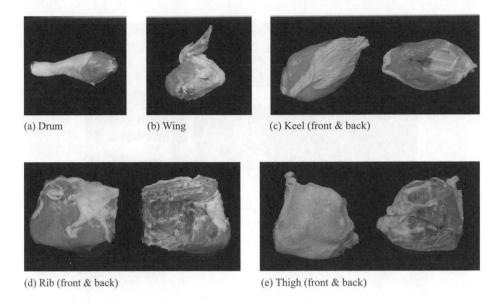

(a) Drum   (b) Wing   (c) Keel (front & back)

(d) Rib (front & back)   (e) Thigh (front & back)

Fig. 3 Various cuts of chicken

possible to distinguish blood-containing eggs from normal eggs by recognizing subtle color differences, minute blood droplets may be difficult to identify. In addition, it is possible to detect cracks using this method.

Among forest products, *shiitake* and other mushrooms have many **grading** criteria such as cap opening. Figure 5 shows a dried *shiitake* sorting system being developed in Korea. This is the third commercial model developed at Sung Kyun Kwan University[59, 60] which moves the mushrooms at a conveyor speed of 150 mm/s and takes 0.7 seconds to grade one mushroom. Since both sides of the *shiitake* mushroom need to be inspected as shown in Figure 6, it features an ingenious inverter mechanism.

A study on shrimp processing[61] sought to apply image processing-based automation to head removal, leaving the edible body and tail parts of the shrimp

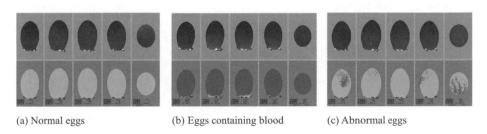

(a) Normal eggs   (b) Eggs containing blood   (c) Abnormal eggs

Fig. 4 Samples of terahertz images (SI Seiko Co., Ltd.)

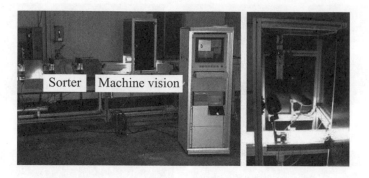

Fig. 5 Dried mushroom sorting system (left) and its machine vision system (right)
(Sung Kyun Kwan University, Korea)

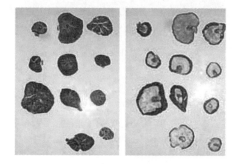   

Fig. 6 Dried shiitake mushrooms   *Fig. 7 Transmittance image of*
(front & back)   *shrimp and binary image of skeleton[62]*

intact. The head cutting position is estimated by extracting the skeletal structure from a **binary image**[1-2.4] of the shrimp (Figure 7) or recognizing the mid-gut gland (hepatopancreas) of the head having a lower density in a transmittance image. Another study[62] set-out to develop a system to recognize the orientation of oysters on the conveyor belt based on the top and side views for the ultimate purpose of shucking them.

2.31 Cattle monitoring

> All homes in the village of Nishiokoppe in Hokkaido are supplied by fiber-optic cables laid down by the Comprehensive Rural Improvement Project of the prefecture which was designated as a model project of the Rural Multimedia Development Program of the Ministry of Agriculture, Forestry and Fisheries. The application of information technology to dairy farming began in 2003 when cattle monitoring robots were installed at eighteen dairy farms to detect problems during calving.

The village of Nishiokkope, Monbetsu County, is located 25 km inland from the Sea of Okhotsk in the northeastern part of Hokkaido Prefecture. This small village of about 1200 residents was chosen as a model district for Hokkaido's Comprehensive Rural Improvement Project, which was designated as a model project of the Rural Multimedia Development Program of the Ministry of Agriculture, Forestry and Fisheries (total budget of 1.7 billion yen). All homes in the village have been connected to an **optical fiber** network, which was constructed as part of the information technology development program in March 2002. Information technology has also been utilized in dairy farming and all eighteen farming households and companies in the village have been fitted with **cattle monitoring robots**.[63] One limited liability company owned by seven people keeps about 350 head of dairy cattle, which means that one calf is born almost every night. Since the outbreak of BSE, recording breeding history (**traceability**) has become an important farming procedure. In the village's system, cattle inventory management software is distributed to participating farms via the Internet and farmers can trace breeding information using a barcode attached to the ear of each cow.

The monitoring robot consists of a monitoring camera, lights, a microphone, a wireless device, a zooming device and a camera turning device. The camera angle, magnification and lighting intensity are remote-controlled. The system has been received well by dairy farmers as they can monitor near full-term cows in the cow shed from their own home, not only watching on a PC monitor but also listening to any sounds the cows make. Figure 1 shows the specifications of the cattle monitoring robot manufactured by Mitsubishi Electric System & Service Co., Ltd. and Figure 2 outlines the remote cowshed monitoring service.

The optic fiber cable can carry a vast amount of data. The village has contracted Kisho Joho Teikyo Co., Ltd. to televise the latest local weather information, which is useful for hay harvesting and other farming operations. The population of the village is decreasing and aging. The system can also be used for providing health care to elderly people living alone who can send their daily blood pressure and

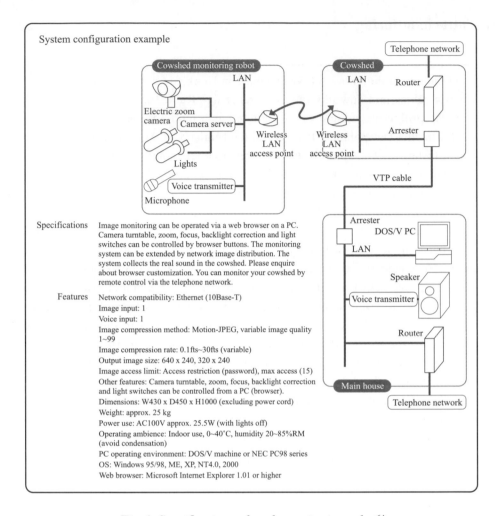

System configuration example

Specifications — Image monitoring can be operated via a web browser on a PC. Camera turntable, zoom, focus, backlight correction and light switches can be controlled by browser buttons. The monitoring system can be extended by network image distribution. The system collects the real sound in the cowshed. Please enquire about browser customization. You can monitor your cowshed by remote control via the telephone network.

Features — Network compatibility: Ethernet (10Base-T)
Image input: 1
Voice input: 1
Image compression method: Motion-JPEG, variable image quality 1~99
Image compression rate: 0.1fts~30fts (variable)
Output image size: 640 x 240, 320 x 240
Image access limit: Access restriction (password), max access (15)
Other features: Camera turntable, zoom, focus, backlight correction and light switches can be controlled from a PC (browser).
Dimensions: W430 x D450 x H1000 (excluding power cord)
Weight: approx. 25 kg
Power use: AC100V approx. 25.5W (with lights off)
Operating ambience: Indoor use, 0~40°C, humidity 20~85%RM (avoid condensation)
PC operating environment: DOS/V machine or NEC PC98 series
OS: Windows 95/98, ME, XP, NT4.0, 2000
Web browser: Microsoft Internet Explorer 1.01 or higher

Fig. 1 Specifications of cattle monitoring robot[64]

pulse data to community nurses via the videophone. The system offers various other services such as the village's CATV service, an agricultural management service which provides sales and material purchasing data for each farm household (from the database accumulated by JA Okhotsk Hamanasu Nishiokoppe Office), an agricultural information service which provides cattle inventory information (basic, milk volume, breeding, disease and raising data) for each farm household, the internet and a resident bulletin board service.[63] Thus, the whole village is connected by **LAN**, including the village administration office, public facilities, schools, and agricultural cooperatives as well as the CATV studio, ordinary households and dairy farms.

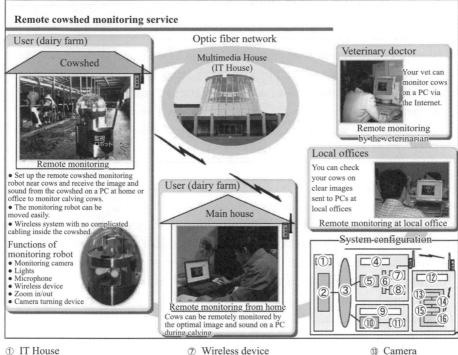

① IT House
② Optical communication center
③ Optic fiber network
④ Dairy farm (home)
⑤ Optical communication terminal
⑥ Hub
⑦ Wireless device
⑧ PC
⑨ Local offices (JA etc.)
⑩ Optical communication terminal
⑪ PC
⑫ Dairy farm (cowshed)
⑬ Camera
⑭ Lights
⑮ Microphone
⑯ Wireless device

Fig. 2 `Remote cowshed monitoring service[64]

2.32 Applications of Terahertz imaging

> Research on terahertz imaging technology is advancing rapidly. Since the terahertz wave has good transmitting properties as well as the ability to provide spectral information inherent to various materials, there are good prospects for applications in the agricultural industry such as plant physiology, vitamins, sugar and other component testing, food testing, agricultural chemical testing and safety testing.

Terahertz waves[1-2.1] are defined as the electromagnetic waves in a spectrum between the far-infrared and microwave regions with a frequency of 10^{12} (THz) and a wavelength of several hundred μm. This spectrum is characterized by the 'ease of handling of light waves,' which allows the use of optical components such as lenses and mirrors, as well as the 'transmitting properties of the radio waves' which are used in mobile phone technology. Despite these attractive properties, the terahertz spectrum has been a largely unexplored territory due to the unavailability of easy-to-use light sources. Some compact and easy-to-operate light sources have been developed in recent years, which have sparked more research in many fields, including medical and solid-state sciences.

Figure 1 presents a schematic diagram of an imaging system using a terahertz wave source; a TPO (**Terahertz wave Parametric Oscillator**)[64] which can freely change wavelength in a broad band between about 1 THz and 3 THz (300–100 μm wavelength) by turning the rotary stage on which a **nonlinear optical crystal** is mounted. The system depicted in this diagram is a transmission imaging system using the light source to directly detect the transmission intensity of the terahertz waves by scanning each pixel. An alternative imaging method was proposed by the research team at Rensselaer Polytechnic Institute, U.S.: a combination of the **Pockels effect** Terahertz field detection method[65] using an **electro-optic crystal**

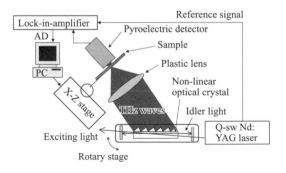

Fig. 1 Terahertz imaging system (RIKEN)

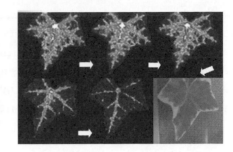

Fig. 2 Terahertz images of tomato leaf

Fig. 3 Water content change in leaf on Terahertz images

as the imaging plate and a **CCD camera**.[1-2.1] This method is very appealing in that it can acquire real-time images, but it is expensive at this point in time.

Figure 2 shows a Terahertz **transmittance image** of a tomato leaf acquired by the system depicted in Figure 1. The tomato leaf was held between two plastic sheets. The wavelength used was 200 μm (1.5 THz) which can capture an image with a spatial **resolution** of several hundred μm. The scanning resolution of this image was 200 μm per pixel and the field of vision was 55 mm x 45 mm. The terahertz wave can achieve this level of clarity of transillumination images because they are less prone to being scattered by the plant's internal or surface structure because it has a longer wavelength than the infrared light. The terahertz wave is also known for its sensitivity to water, a characteristic that can be used for non-destructive internal **moisture monitoring** of plants.[66] Figure 3 shows the result of moisture monitoring by imaging of moisture change in a leafy plant. Substantial changes in moisture distribution over time can be observed. Visualization of moisture levels previously required equipment such as **MRI**[1-2.5] but the availability of Terahertz wave sources is expected to contribute to the field of plant internal moisture research. Figure 4 presents examples of Terahertz images.

(a)	(b)	(c)	(d)
Cherry petals and leaf	Dried shrimp	Red pepper	Needle and cutter blade in cotton

Fig. 4 Samples of terahertz images

The characteristic absorption spectra of vitamins, agricultural chemicals and sugar has been discovered within the terahertz spectrum, which has led to the development of technology for material identification and measurement using **spectroscopic imaging**.[67] Also, a difference in absorption between the single-stranded and double-stranded forms of DNA in the Terahertz band has been confirmed;[68] the application of this finding to label-free gene analysis is anticipated.

2.33 Applications of Hyperspectral imaging (AOTF)

> New techniques have been developed for acquiring three-dimensional information by adding the spectral information of each component and the composition of the target object to two-dimensional spectrum information. These are called multi-spectral imaging and are applied to remote sensing. This section describes hyperspectral imaging using AOTF which uses even higher wavelength resolution and can acquire images at high speed.

Figure 1 is a schematic illustration of a **multi-spectral image**. A multi-spectral image is a collection of images with different wavelengths, as illustrated, which are usually acquired in three or more bands. In other words, it contains three-dimensional information of spatial distribution along the X, Y and wavelength axes. Analysis of such images can highlight aspects of the target object that are difficult to distinguish in images such as photographs. Hence, this type of system is widely used on satellites and airplanes as a means of **remote sensing**. In particular, images having a high wavelength resolution of several nm are called **hyperspectral images**,[1-2.5] which are attracting attention not only in the field of environmental mapping but also in the medical and cultural asset survey fields.

The multi-spectral image is usually acquired by an image pickup device such as a CCD fitted with a wavelength filter using only the light in the band that passes through this filter. The filter is replaced with another one to produce another image in a different wavelength. This process is repeated to produce an image in multiple spectrums. In recent years, however, the appearance of various optical devices has made acquiring hyperspectral images with even higher wavelength resolution easier and quicker. Figure 2 is a schematic diagram of hyperspectral imaging using a filter called **Acousto-Optic Tunable Filter** (AOTF). The AOTF involves tellurium dioxide (TeO_2) crystals which allow light of a certain range of wavelengths to pass through as their refractive index changes when vibrations of a certain frequency are induced. By controlling the frequency applied to the crystals, they can be used as filters for a given wavelength range. Such filters are included in optical systems which are marketed by BRIMROSE in the U.S. and several other manufacturers. These AOTF units have no moving parts, are easily controllable, and can be controlled at high speeds of several milliseconds/wavelength. The images thus acquired are called hyperspectral images rather than multi-spectral images and have been attracting interest in recent years. Highly sensitive cooled CCD cameras are often employed for this method since the underlying principle of the method means that only a faint light reaches the CCD through the AOTF.

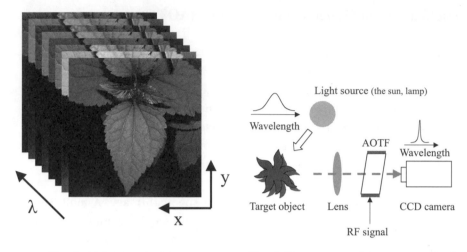

Fig. 1 Multi-spectral image

Fig. 2 Hyperspectral imaging method
using AOTF

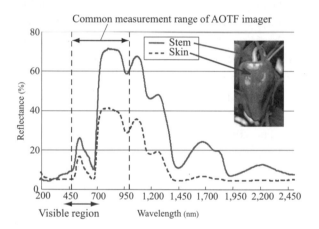

Fig. 3 Spectral reflectance of sweet pepper

Figures 3 and 4 present examples of an application of spectral imaging. Figure 3 shows the **spectral reflectance**[1-2.2] properties of the stem and skin of a sweet pepper. Due to a large difference in reflectance in the near infrared region outside of the **visible region**,[1-2.1] it is possible to discriminate the stem and the skin, which appear to have the same green color, without using any shape-related information by computation processing of the acquired spectral image (Figure 4). This image separation technique can be used for the vision unit of agricultural robots and has prospects for applications in harvesting work and crop management.

Many research facilities are currently working on methods for using this technology to estimate the rate of photosynthesis and the amount of

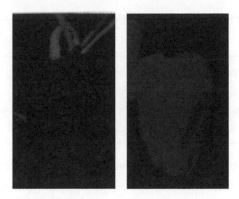

*Fig. 4 Discrimination between fruit
and stem of sweet pepper using
multi-spectral information*

chlorophyll.[69,70] They have successfully estimated the rate of photosynthesis and visualized the distribution of chlorophyll by analyzing vast amounts of acquired spectral images for the purpose of environmental mapping by remote sensing. Another development is a method for simultaneous material identification and quantification by analyzing the degree of similarity between the acquired image and the previously measured spectrums using image analysis and principal component analysis techniques.[71]

2.34 Applications of remote sensing

> Remote sensing methods include ground-based remote sensing using vehicles, aerial remote sensing from low- to medium-flying aircraft such as unmanned industrial helicopters, and satellite-based remote sensing from space satellites. Each has its own strengths and weaknesses.

The topical issue in the field of **Precision Farming** (PF) is how to apply remote sensing technology to the assessment of the growing conditions of plants and how to reflect the acquired information in precision farming management. Ground-based **remote sensing** from vehicles and satellite-based remote sensing using **satellite images** to survey the growing conditions of plants are attracting interest. The information acquired by remote sensing can be analyzed to produce yield estimates and quality information such as protein content. In recent years, very high **resolution** satellite imaging systems such as IKONOS and QuickBird have come into practical use, raising hopes for their application as sensing devices for food production.

For example, satellite images can be used to create a sectional map of soil organic matter content as in Figure 1 (see CD-ROM). Pink signifies areas with 5% or less organic matter content which tend to have low soil moisture as well. Green and yellow signify areas with higher organic matter content in the soil. Analysis of high resolution images can clearly demonstrate different soil conditions within a field. Since the soil organic matter content is linked to the amount

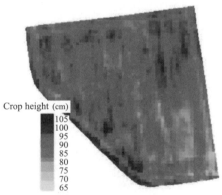

Crop height (cm)
105
100
95
90
85
80
75
70
65

Fig. 1 Sectional map of soil organic matter content (Chiba University)

Fig. 2 Crop height map of wheat field (NARC Hokkaido Region)

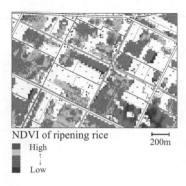

NDVI of ripening rice   ⊢———⊣
■ High   200m
  ↑
  ↓
■ Low

Fig. 3 Normalized Difference Vegetation Index
(NDVI) map of rice paddies (Hokkaido Central
Agricultural Experiment Station)

of hot water-extractable nitrogen, this method can be used for soil fertility assessment. It can also be used in land improvement projects for poorly drained areas and determination of crops appropriate for the soil condition. Figure 2 is a map of crop height in a wheat field using QuickBird imagery. Variations in wheat growth within the field can be observed. Figure 3 is a **Normalized Difference Vegetation Index** (NDVI) map of paddy fields of ripening rice using IKONOS images. Because the NDVI during the ripening stage has a high correlation with the protein content of rice grain, this map can be considered to be the equivalent of an eating quality map. In Hokkaido, protein content maps are stored in the agricultural GIS (Geographical Information System) of each town. Images for particular years and areas can be viewed and output in enlarged format on demand. The information is combined with data provided by cooperatives and available for detailed analysis such as sorting by farm owner.

Figure 4 presents mosaic images of a corn field acquired by an **unmanned industrial helicopter**. Small images are converted into ground coordinates and pasted on the GIS one by one. Naturally, spatial **resolutions** of these images are far higher than those of satellite images; the image in Figure 4 has a ground resolution of 3 cm. The map in Figure 5 was generated based on NDVIs calculated from this image. The NDVI information can be used for precision management work such as variable topdressing application due to its correlation with leaf nitrogen content. Figure 6 shows soil organic matter content maps generated from the IKONOS images and the images taken from a low-flying unmanned industrial helicopter respectively. Both imagers are fitted with vision sensors capable of acquiring red and near infrared images and are therefore able to generate similar soil organic matter maps. These results suggest that hybridization of satellite and helicopter platforms can lead to the construction of a new remote sensing systems.

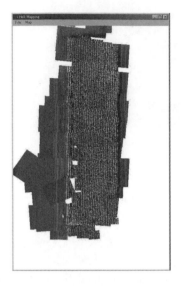

*Fig. 4 Mosaic field images acquired
by unmanned helicopter*

*Fig. 5 NDVI map generated from
images acquired by unmanned helicopter*

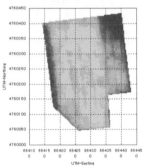

IKONOS satellite image

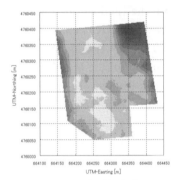

Image acquired by helicopter

Fig. 6 GIS maps of soil organic matter content

Coffee break

Why is the sunset red?

Many of you would see a bright red sky at sunset on your way home and think about fine weather the following day. You don't usually wonder why it is red unless your children ask. As you can observe with a rainbow or a prism, sunlight can be separated into many different colors and, as you know, red has the longest wavelength and violet has the shortest wavelength among visible colors. However, most people are unaware that when they look toward raindrops floating in the air with the sun at your back, a rainbow is formed at a 42-degree angle due to refraction and reflection of sunlight (refractive index of water is 1.337). Sunlight passes through the earth's atmosphere before reaching the ground. When the sun is at its highest point in the sky at around noon as in the diagram below, light passes through a relatively thin atmospheric layer (shorter travel distance). It passes through a thicker layer (longer travel distance) at sundown. The sky appears blue because short-wavelength light is scattered off molecules in the atmosphere and long-wavelength light is transmitted. The atmospheric scattering of light is greater at short wavelengths than long wavelengths (Rayleigh scattering which is inversely proportional to the fourth power of wavelength). Although short-wavelength light is still scattered at sunset, it cannot travel through the thick atmospheric layer. Only long-wavelength light (red light) appears to reach the human eye and create the evening glow effect. For reference, light reaching the ground is often expressed in terms of color temperature according to different color components. For example, it is between 2000 and 2500 K at dawn and sunset, between 5000 and 5500 K at around noon on a fine day and around 6500 K on a cloudy day. Light is reddish at lower color temperatures and bluish at higher color temperatures. *(N. Kondo)*

3 End-Effectors and Arms of Agri-Robots

The end-effectors of agri-robots are purpose-built to suit the target objects and tasks. The basic mechanism of the robot arm is determined by the cultivation system used to grow the target object. Tasks that are usually performed by the two arms and ten fingers of a human being are often performed by agri-robots more rationally using totally different methods. This chapter will describe the end-effectors and arms of agricultural robots with various innovative ideas.

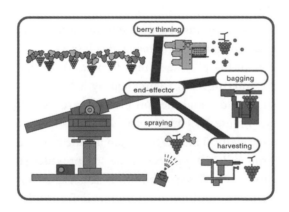

3.1 Seedling transplanting (Movie)

> For transplanting plug seedlings, the development of a sliding needle type robot at Rutgers University, U.S., was reported in 1990. Known commercial models include the potting robot by YANMAR Co., Ltd. and the complementary transplanting system by Visser International Trade & Engineering B. V., Holland.

Among various types of transplanting methods for seedling production, this section will describe the **plug seedling** method that can be easily mechanized. Rutgers University, U.S., has developed a robot which transplants small **cell tray**-raised seedlings to larger cells.[1] The left side of Figure 1 shows the end-effector[2] [1-3.5] of this robot. This is called a 'sliding needles' gripper which inserts two needles into the cell at an angle, then lifts and carries the seedling to another cell like a pair of chopsticks. In this way, it can transport seedlings more smoothly than the human hand does without gripping the seedling itself (Movie 1). The arm is a commercially available **SCARA type**[1-3.6] arm (four-degrees-of-freedom **Selective Compliance Assembly Robot Arm**) (Figure 1 right). The application of machine vision[1-2.1] to complementary transplanting was also trialed in an attempt to transplant seedlings to selected cells based on an acquired image of cell trays.[3]

Robots for transplanting plug seedlings to pots have been commercialized. Figure 2 shows the **potting robot** launched by YANMAR Co., Ltd. in 1997 (Movie 2). The pot provider and the soil provider are used in earlier stages of this potting process (Figure 3). The pot provider uses suction to feed four plastic pots to a tray automatically. This machine can handle pots with diameters of 75 mm, 90 mm, 105 mm and 120 mm. Next, the soil provider shakes the pots and fills them with culture soil. Then, the potting robot uses an end-effector with two spade-shaped claws to pull out some seedlings from a 512-hole or 406-hole

Fig. 1 Transplanting end-effector and arm[1]

Fig. 2 Potting robot   Fig. 3 Pot provider and soil provider

cell tray and uses a **Cartesian coordinate arm**[1-3.6] to move them to a conveyor belt. The seedling sensor on the conveyor belt detects the presence of seedlings and identifies defective seedlings. An end-effector with four needles attached to a Cartesian coordinate arm transplants only the acceptable seedlings to the pots. The number of simultaneously operating end-effectors is determined based on the number of pots in this transplanting work. This robot system is capable of processing 6000 seedlings per hour.

Visser International Trade & Engineering B. V. of Holland has also commercialized various types of seedling production machines.[4] One of them is a robot which performs complementary plug seedling transplanting as in Figure 4. This machine detects defective seedlings with **CCD camera**-based machine vision[1-2.1] and uses two end-effectors to remove defective seedlings (right end-effector) and transplant acceptable seedlings (left end-effector) simultaneously (Movie 3). This system has a processing capacity of 3000 seedlings per hour. Another system which inspects seedlings with a CCD camera and transplants acceptable seedlings to cell trays using two end-effectors can process 4500 seedlings per hour (Figure 5). The other type which blasts pressurized air to the bottom of the tray to remove defective seedlings has the same transplanting capacity.

Fig. 4 Complementary   Fig. 5 Machine vision system (left) and
transplanting system[4]   transplanting end-effector (right)[4]

3.2 Chrysanthemum cutting sticking (Movies)

> Joint research on a chrysanthemum cutting sticking robot system was launched by Okayama University, Panasonic Corporation and JA Aichi in 1993 and produced a prototype model. BRAIN and ISEKI & Co., Ltd. have jointly developed a cell tray cutting sticking machine as part of the Urgent Project 21.

Although direct cutting sticking into the ground has been widely practiced in Japan's chrysanthemum culture, the use of **plug seedlings** in cell trays has become common in recent years for improved root-taking, seedling uniformity and ease of handling. Figure 1 is a diagram of the work procedure. First, an axillary bud (cutting) that has grown to an acceptable size is picked from the parent stock and stored. Once a sufficient number of cuttings have been picked, they are conditioned for water uptake, trimmed of lower leaves and planted in cell trays. Transplanting machines are already available on the market for seedling transplanting in the next stage but the cutting sticking procedure is yet to be mechanized and hundreds of millions of cuttings are still being processed by hand each year. Figure 2 shows the conventional cutting sticking operation and Figure 3 shows chrysanthemum cuttings.

The **cutting sticking robot** system is outlined in Figure 4 (Movie 1). This system is comprised of a separator & feeder[5] which feeds conditioned cuttings one by one, a leaf remover[6] which removes excess lower leaves, a cutting sticking device[7] which plants cuttings in cell trays, an **articulated arm**[1-3.6] which transports cuttings, and a vision system[8, 9 (II-2.2)] which detects the orientation and position of each cutting.

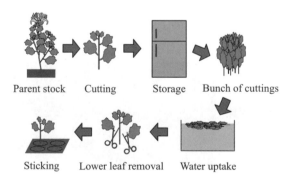

Parent stock   Cutting   Storage   Bunch of cuttings

Sticking   Lower leaf removal   Water uptake

Fig. 1 Chrysanthemum cutting sticking procedure

Fig. 2 Conventional cutting sticking
operation

Fig. 3 Chrysanthemum cuttings

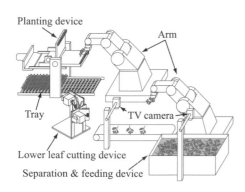

Fig. 4 Cutting sticking robot system
(Okayama University)

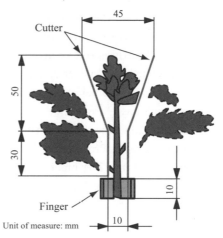

Fig. 5 Y-shaped cutter

In manual operation, workers nip lower leaves at the axil with their fingers, leaving a total of two to three leaves on each cutting before sticking it into a tray. Leaf removal is done in order to facilitate sticking cuttings into a tray and transplanting by a **transplanting machine**. It also minimizes moisture evaporation and improves aeration to prevent disease. With the current technology, this action of nipping individual leaves by the fingers is difficult for a robot to emulate and perform accurately. It is very difficult for the vision system to discriminate individual leaves from dense foliage since chrysanthemum leaves tend to spread in all directions. Removal of such leaves would require complex mechanism and control. For these reasons, this robot uses a cutter mechanism[1-3.5] to remove excess leaves at once. The leaf remover is mainly comprised of a fixed Y-shaped cutter[1-3.5] and opening-closing flat plates with cushioning material. When the arm brings a cutting to position, the plates close and the cutter removes its outer leaves (Figure 5). A second leaf remover is placed at a 90 degree angle to the first unit so that large leaves that cannot be removed fully from one direction can be cut by the second unit (Figure 6). The positioning and dimensions of the cutter were determined on the basis of research on the form and size of actual cuttings. Figure 7 (right) shows a cutting after leaf removal.

The trimmed cuttings are laid one by one on the planting device by the robot arm (Figure 8(a)). When ten cuttings are laid on the device, the cover closes (Figure 8(b)), the device descends as it rotates downward and sticks a row of cuttings in the cell tray (Figure 8(c)). This rotating and descending action of the planting device enables it to stick cuttings in targeted cells unobstructed by the previously planted cuttings.

Figure 9 shows the prototype which was trialed for practical application based on the above research outcome. The system uses a camera to recognize manually fed cuttings and an arm to move them to the conveyor. A cutter moves up and down to remove excess leaves on the conveyor, then the cutting is turned 90 degrees and the other cutter removes some more leaves. The arm plants the cutting in the tray (Movie 2).

Figure 10 shows the **cutting sticking machine** for **cell trays** developed by the 21st century Urgent Project and Figure 11 shows the cutting sticking operation using this machine. The machine consists of the cutting feeder, the lower leaf processor, the lower leaf remover, the cutting transporter and the cutting sticking unit, and plants ten cuttings at a time. The flow of operation is as follows. The chrysanthemum cuttings that have been prepared to an appropriate length and soaked in a rooting stimulator are fed into funnel-shaped feeders by the operator by hand. The system adjusts the lower end and posture of each cutting and moves them to the lower leaf remover. Each cutting is inserted into a space between two pairs of revolving brushes, which remove the lower leaves by their rotating action. The trimmed cutting is positioned so that its lower end comes to the center of each cell and is moved to the cutting sticking unit. The cell tray, which has been filled with the culture soil and watered, is fed into the cutting sticking unit by the operator. The row by row sliding action of the cell tray is linked to the planting action of the sticking unit. In the stage prior to planting, holes are made in the soil-filled cells. There are twenty cells in a row and ten cuttings are planted in every other cell in each planting action, which means that the planting of one row is completed in two operations. Figure 12 shows a cell tray planted by this machine immediately after planting (Movie 3).

Lower leaf removal experiments on the standard type, spray type and small-flowered chrysanthemum cultivars have found that the most practical settings for the machine included a processing time of 3–4 seconds for standard chrysanthemums and 2 seconds for spray and small-flowered chrysanthemums,

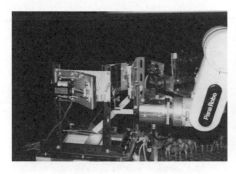

Fig. 6 Lower leaf cutting device   Fig. 7 Cuttings before and after lower
                                                      leaf removal

Fig. 8 Planting device

Fig. 9 Prototype cutting sticking robot system
(Panasonic Corporation)

a brush rotation rate of 80 rpm, and a stem length for leaf removal by the brushes
of 0–20 mm adjusted according to the length of each cutting. Performance
tests using two cultivars of standard chrysanthemum, three cultivars of spray
chrysanthemum and four cultivars of small-flowered chrysanthemum have
resulted in misplanting rates ranging from 0.3 to 6.6 percent and more uniform

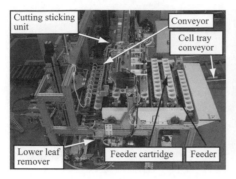 

*Fig. 10 Semi-automatic cutting sticking Fig. 11 Chrysanthemum cutting feeding
machine (BRAIN, ISEKI & Co., Ltd.)   operation*

planting depth than manual planting. Even when the cuttings were transplanted in slightly shallower positions than the set value of 25 mm, they grew well both above and below the soil as shown in Figure 13. The work rate of the machine with one operator was 1788–2158 cuttings/hour, which was about twice the speed of the conventional manual operation.

Another type of machine has been developed for seedling producers with cutting production bases in overseas countries. It plants cuttings which have already been trimmed of lower leaves. It has four adjustable planting speeds between 1500 and 5200 cuttings/hour and can be operated by up to two people. It has a misplanting rate of 0.5–2.7 percent and a single-operator work rate of 1900–3030 cuttings/hour. This type of machine with improved operability was brought to the market in 2005.

*Fig. 12 Plug tray with planted
cuttings*   *Fig. 13 Seedling root growth*

A grafting robot for cucumbers launched in 1993 was the first 'grafting robot' for the industry. Under the Agricultural Mechanization Promotion Law, BRAIN developed grafting robots for cucurbits and *Solanum* species in 1994 and 1995 respectively as part of the Urgent Project. Fully automated models have been commercialized.

Grafting has been practiced widely for a very long time as a technique that is indispensable for the prevention of crop failure from continuous cropping and the stable production of high quality fruits. The volume of grafted nursery plant production in Japan is approximately 400–600 million seedlings per year. A majority of commercially available seedlings of tomatoes, eggplants, cucumbers, and melons are grafted. While grafting has traditionally been done by the hands of experienced and skilled workers, the common problems of the farming industry in recent years, i.e., aging workers and labor shortage, are forcing farmers to rely on commercial operators and cooperative nursery centers for seed production. This situation has increased the need for labor saving and cost reduction in mass seedling production. Against this backdrop, BRAIN and five private companies have jointly developed and commercialized a **semi-automatic grafting robot** for the mass production of grafted seedlings.

The operator manually feeds a scion and a rootstock into the semi-automatic grafting robot, which cuts their stems and joins them together with a clip to produce a grafted seedling. This simplified description may suggest that it is a rather simple system, but in fact individual plants of any species can have various shapes, sizes, hardness, curvature and deformities. Various ingenious ideas have been incorporated in the robot to deal with these challenges and achieve secure union and graft-take. Figure 1 shows the semi-automatic grafting robot[10] and Figure 2 shows the operators using it. The robot is comprised of a symmetrically positioned feeder unit, a grip and transport unit, a cutting unit, and a joining unit at the center which joins the scion and the rootstock with a clip. The operator can put clips in the parts feeder in the back of the joining unit in no particular order; the machine aligns them and feeds them into the system one at a time. All of the working parts except for the parts feeder are pneumatically actuated and continuously perform a sequence of basic actions, including feeding, gripping, transporting, cutting, transporting, joining and ejecting, under the **sequential control** of the programmable controller.

The system can be operated easily by beginners as scions and rootstocks fed into the respective feeders are automatically grafted by the machine. When plants

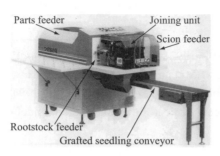

Fig. 1 Semi-automatic grafting robot
(ISEKI & Co., Ltd., BRAIN)

Fig. 2 Operators using semi-automatic
grafting robot

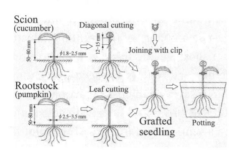

Fig. 3 Leaf cutting method of grafting

Fig. 4 Fully-automatic grafting robot
(YANMAR Co., Ltd.)

are detected by the sensors fitted to the scion and rootstock feeding tables, the gripper/transporter devices carry them to the cutting devices by rotating them 90 degrees. Excess parts of the plants are then cut off by razor blades. The cut plants are carried to the joining device by another 90 degree rotation and clipped together. This sequence of actions completes one grafting cycle. Each cycle takes approximately 4.5 seconds and the system is capable of producing 800 grafted seedlings per hour in continuous operation. Since the system does not perform grafting until plants are fed to the scion and rootstock feeders, it works according to the feeding speed of individual operators (Movie 1).

For grafting cucurbits, a leaf cutting method (Figure 3) has been developed and employed rather than simple automation of the conventional hand grafting method for better mechanical handling and rates of successful union. This method entails cutting one leaf and the growing point of the rootstock at an angle and cutting the embryonic axis of the scion at the same angle. The two cut surfaces are joined and fixed with a special clip. Precision cutting is required for the rootstock since the growing point must be completely removed together with one leaf. The developmental base of the cotyledon of pumpkin seedlings, which are widely used as rootstocks for cucurbits, has a uniform shape regardless of seedling size and the system can perform highly accurate and fast operations using a simple mechanism by aligning this part with the upper side of the seedling feeder unit. It would be no exaggeration to say that thorough research on the **physical properties**[I-3.2] of the target object has led to the practical application of the grafting robot.

For *Solanum* species, both semi-automatic grafting robots modified from the cucurbit grafting robot and fully automatic grafting robots have been commercialized. Figure 4 shows a **fully-automatic grafting robot**[11] (Movie 2). Scions and rootstocks raised in **cell trays** are fed into this robot as they are. After the diagonal cutting of the embryonic axis of both plants and joining, grafted seedlings are planted in empty seedling trays and discharged from the robot. Although it is a fully automated system, it cannot process all types of seedlings. Robot-ready seedlings must be raised to satisfy certain criteria in order to maintain a consistently high grafting success rate. Once this hurdle is cleared, however, the robot can process one cell tray row of seedlings at a time and produce grafted seedlings at a rate of 1000–1300 seedlings per hour and a success rate of over 95 percent.

3.4 **Pruning** (Movie)

> Pruning involves cutting off tree branches up to a certain height for the main purpose of producing high quality timber with no or few nodes. Commercially available pruning machines have wheels, which clasp and turn around the tree trunk. Pruning machines cut branches off as they climb up the tree in a spiral motion. Basic research is also underway on a straight climbing mechanism which moves in a manner similar to an inchworm was also introduced.

Forestry work involves operations such as ground clearance for planting seedlings, planting, weeding and brushing, cutting out excess timber growth, thinning, harvesting, transporting and pruning. While various machines have been developed, including chainsaws, **brush cutters**, processors for delimbing and cutting (bucking), yarders, winches and hauling machinery, many heavy physical operations still have to be performed by human workers due to rugged or steep ground surfaces in the mountains. Pruning involves cutting off tree branches from the base up to a certain height, mainly for the purpose of producing high quality timber with no or few nodes. It is a dangerous operation at heights and requires skilled workers. It has been neglected in recent years due to depressed timber prices, increased timber imports and labor shortages. Hence, self-climbing pruning machines (**pruning robots** or automatic pruning machines) are being used for the sake of operational safety and labor saving.

Figure 1 shows a pruning machine that prunes tree branches as it climbs a tree in a spiral motion (Movie). The upper and lower parts of the machine each have a set of five diagonally positioned wheels which clasp the tree trunk and exert frictional forces using extension springs. The machine climbs up and down in spirals by rotating these wheels. A chain saw is fitted to the top of the machine in close proximity to the tree trunk so that it can cut branches at the base as the machine climbs up in spirals. It has an overload protection function which shifts the engine down to low revolution when the chain saw becomes jammed or overloaded during pruning operation. It also has a jam reduction mechanism by which the chain saw retreats slightly when the bar at the top of it hits a branch.

The machine is attached to a tree with two extension springs as if it is wrapped around the tree trunk. It begins pruning operation once its engine is turned on. When the machine reaches a designated height, the operator on the ground sends a stop signal from the radio controller. It uses a gasoline engine with a maximum output of 1.4 kW, a pruning and ascending speed of 1.8–2.5 meters/min and a descending without pruning speed of 5.9–8.9 meters/min.

Figure 2 shows the inchworm-like straight climbing mechanism,[12] which was test-manufactured as part of basic research for: ① selective pruning, ② reduced

Fig. 1 Pruning machine (Seirei Industry Co., Ltd.)

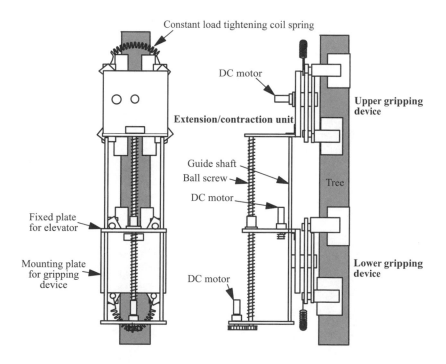

Fig. 2 Climbing mechanism (Courtesy of Yoshihiko Takimoto)

compression forces exerted by the elevator on the tree trunk to prevent damage to the cambium, and ③ weight reduction for improved portability.

This climbing mechanism has upper and lower gripping devices that grip the tree and an extension/contraction unit that links the two sets of gripping devices. It is driven by a ball screw and motors, which open the upper grips, close the lower grips to hold the tree, and extends the extension/contraction unit to elevate the upper grips. It holds the tree with the upper grips, opens the lower grips, and contracts the extension/contraction unit to elevate the lower grips. The machine ascends by repeating these actions and descends in a similar manner. The gripping devices and the extension/contraction unit are driven by **DC motors**[1-3.3] and are able to move up and down automatically using an 8-bit microcomputer. The combined weight of the gripping devices and the extension/contraction unit is 11.6 kg.

> Research to develop a multipurpose robot for grapes growing on trellises began at Okayama University in 1989. It can perform harvesting, berry-thinning, and chemical application as well as grape bagging by using different end-effectors.

Trellis training[(I-3.2)] on horizontal trellises at around the height of an adult is used to grow a large proportion of grapes in Japan (Figure 1). This cultivation method was devised to reduce the risk of diseases caused by humid summer climate. However, it gives rise to demanding work conditions for vineyard workers who are forced to prune, thin, bag and harvest with their heads tilted back and arms elevated for many hours. Figure 2 shows workers who are treating grapes with gibberellin (used to produce seedless grapes).

Figure 3 shows the **grape harvesting robot**[13] which performs harvesting and various crop management procedures for these trellis trained grapes. An end-effector and a vision device are attached to the end of the robot arm, which is mounted on the **crawler** type traveling unit. It is one of the multipurpose robots that can perform multiple tasks with various types of detachable end-effectors. With the trellis training system, the arm does not have to have a high degree of freedom since the fruits are distributed on a plane parallel to the ground and there are few obstacles around them. Because the end-effector tends to approach the fruit horizontally or from below, an arm equipped with a **prismatic joint**[(I-3.6)] can perform tasks quickly with relatively simple control. This robot has a five-degrees-of-freedom **polar coordinate arm**[(I-3.6)] incorporating one prismatic joint.

Fig. 1 Trellis training system of grape growing

Fig. 2 Crop management operation

Fig. 3 Multipurpose robot for
grapevines

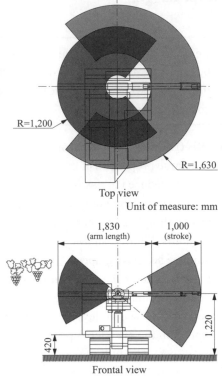

Fig. 4 Operational space of arm

Figure 4 shows the **operational space**[I-3.6] of the arm and Figure 5 describes the basic mechanism of the arm. The arm has four rotational joints[I-3.6] and one prismatic joint, all of which, except for Wrist 2, are driven by the DC servo motor. Reduction gears are fitted to the output shafts to generate higher torque. The revolutions of the reduction gear's output shafts are transmitted to the rotational joints via a chain and a belt. The translation unit (prismatic joint) has a slit on its side and the rotation of the gearwheel is converted to a linear motion by the rack fitted inside the translation unit, which moves the whole translation unit forward and back in a sliding motion. The **DC motor**[I-3.3] shaft of Wrist 2 is fitted with an end-effector. The whole section can be rotated up and down by Wrist 1. The Wrist 1 joint is used to move the end-effector closer to the fruit horizontally or from below and the Wrist 2 joint is useful when harvesting obliquely hanging clusters. This arm and the traveling unit are also used with the **grape harvesting end-effector**[I-3.7] and the **berry-thinning end-effector**.[I-3.6]

Bagging is an important task in grape production in Japan. The aims of bagging include protecting grape clusters against rain, chemicals, splitting, sunscald and

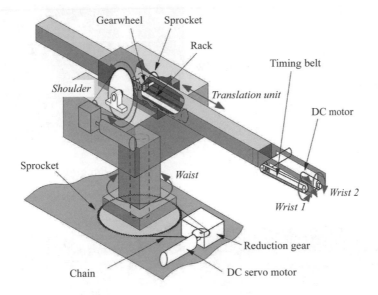

Gearwheel   Sprocket

Rack

Timing belt

Shoulder   *Translation unit*

DC motor

Sprocket

Waist

Wrist 2

Wrist 1

Reduction gear

Chain   DC servo motor

Fig. 5 Basic mechanism of arm

bird/insect damage and controlling the timing of fruit coloring and harvesting as shown in Figure 6. Depending on the variety of the fruit and the production aims, various types of bags are used, which are generally made of moisture resistant machine glazed poster paper or waxed paper. There are both open bottom type bags and closed bottom type bags.

The bag is placed over a grape cluster and a wire tie attached to the top opening of the bag is fastened securely to the rachis (Figure 7). While this is an important operation for the production of high quality grapes, it is a physically demanding task for workers who have to bag bunches one by one while elevating their arms over their heads for many hours.

To have a robot perform this task, the use of wire ties is not practical, as tying a wire would require a highly complex end-effector mechanism and control. A simpler method to achieve the same effect as manual tying needs to be devised. Also, the end-effector needs to have a continuous bag feeding function since it is inefficient for the robot to fetch a new bag for each cluster.

A **bagging end-effector**[14] (Movie) and a robot fitted with it are shown in Figures 8 and 9 respectively. The bag used for robot bagging operations is shown in Figure 10. The end-effector is comprised of open-close fingers to hold a bag and a mechanism to supply bags continuously to the finger. The bag opening has a set of plate springs instead of a wire tie which opens when force is applied from the two ends and closes when released. With the fingers open (Figure 11(a)), the

To protect grape clusters against:
□ Disease transmitted by rain water
□ Contamination by chemical spray
□ Sunscald
□ Bird and insect damage
□ Fruit splitting

Fig. 6 Aims of bagging   *Fig. 7 Bagging operation*

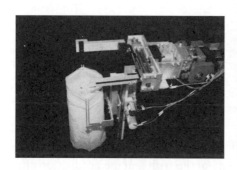

Fig. 8 Bagging end-effector   *Fig. 9 Grape bagging robot*

Plate springs

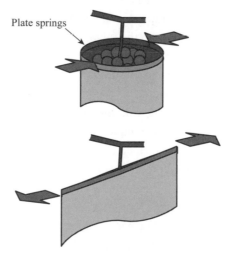

Fig. 10 Bag with plate springs

feeder device with multiple bags ascends (Figure 11(b)). The fingers close and hold a bag with its top open. The bag feeder descends and the arm approaches a bunch of grapes from directly below and slips the bag over it (Figure 11(c)). The fingers open and release the plate springs, which close the bag naturally to complete the task.

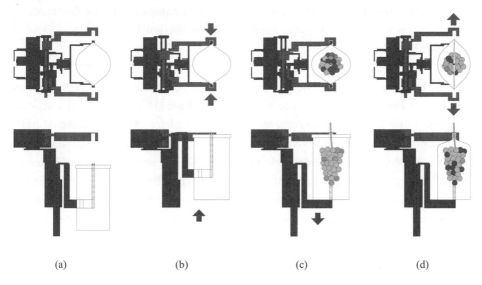

(a)   (b)   (c)   (d)

Fig. 11 Bagging procedure

3.6 Grape berry-thinning

> The development of an end-effector for the multipurpose robot for trellis-trained grapes to simultaneously perform rachis trimming and berry-thinning began in 1989 at Okayama University. The robot can be used for other tasks such as harvesting and bagging (as described in section 3.5) by replacing the end-effector.

Grape clusters contain multiple berries. Crop management therefore often involves individual berries as well as the cluster as a whole. In particular, the cluster trimming and berry-thinning operation is essential for the production of high quality grapes and conventionally involves trimming each cluster by hand. Rachis trimming is done around the flowering period for the purpose of improved bearing, as well as to enhance sugar concentration and coloring, both of which are impeded by cluster overgrowth. Based on consumer preferences and ease of transportation, the cluster is trimmed into a cylindrical shape from its natural inverted cone shape. Berry-thinning is done after fruit setting to limit the number of berries on a cluster to the optimum level in order to standardize the berry size, promote berry growth and prevent fruit cracking from excessive berry setting.

While the method and timing of rachis trimming and berry-thinning vary depending on the cultivar and tree vigor, a conventional method for Kyoho, Pione and Campbell is shown in Figure 1. Shoulders (uppermost branches) and two to four upper rachis-branches are removed with shears around the flowering period. Next, some rachis at the bottom are removed to leave twelve to sixteen rachis-branches (8–10 cm) in the middle. Very long rachis-branches are trimmed so that the cluster forms a cylindrical shape. Within 15–25 days of flowering, berries are thinned by shears to achieve a cluster weight of 300–350 grams at harvesting. The number of remaining berries is between twenty and thirty in the case of Pione and between sixty and seventy in the case of Campbell.[15, 16]

For a robot to perform the abovementioned procedures, it requires a high precision **vision sensor** to recognize individual rachis-branches as well as a highly complex mechanism and controller on the end-effector. Accordingly, the **berry-thinning end-effector**[17] (Movie) shown in Figure 2 has been designed to complete the same tasks using completely different procedures from those used by human workers. First, it trims the upper part of the cluster by removing berries (Figure 3) instead of removing shoulders and rachis-branches. It uses a pair of plates with corrugated polyurethane surfaces to compress a grape cluster and slides them in opposite directions to cause some berries to detach from their peduncles. The stroke between the two plates is set at 100 mm and the compression load is set at

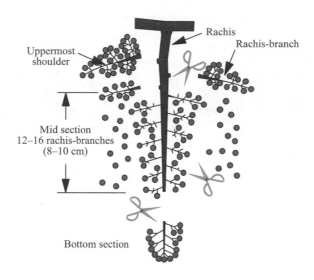

Fig. 1 Conventional method of grape berry-thinning and cluster trimming

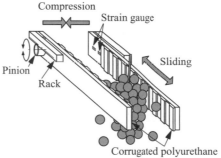

Fig. 2 Grape berry-thinning end-effector   *Fig. 3 Threshing section*

11 N based on the results of preliminary tests. It performs five sets of back and forth slides, then opens the plates by 30 mm to take in more berries while dropping the detached berries, and then repeats this procedure. It uses a **strain gauge**[I-3.4] to constantly monitor the compression force. For berry-thinning in the middle of the grape cluster, needles are used to puncture berries, which later dry up and drop off naturally as shown in Figure 4. The device consists of a plate fitted with several needles, another plate with holes in needle positions and four shafts and springs between the two plates. A pair of these devices squeezes the cluster from either side and the needles protrude and puncture berries once a certain level of force is applied. As shown in Figure 5, it uses two bed knives (plates) and a cutting blade which open and close to sever the bottom part of the cluster.

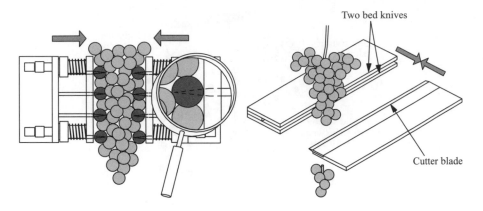

Fig. 4 Thinning section   *Fig. 5 Cutting section*

Grape harvesting

> The development of a grape harvesting end-effector attached to the arm of a robot to work in vineyards began in 1989 at Okayama University. This end-effector performs multiple tasks, including rachis exposure and cluster orientation.

When grapes are harvested manually, the rachis is held in one hand and scissors in the other hand to cut the rachis (Figure 1). Direct hand contact with the berries may cause them to drop, or it may wipe bloom off of the berry surface, either of which reduces the grape's commercial value. When a robot performs this operation, it is still desirable to minimize contact with the fruit cluster while cutting the rachis. The initial data needed for the development of the end-effector include the level of force required to cut the rachis and the level of force required to hold the rachis to prevent dropping the harvested cluster. A very large capacity actuator could cut the rachis properly but it might mean a larger and heavier end-effector than necessary and hence a larger arm to move the end-effector. It is therefore important to understand the **physical properties**[1-3.2] of the target object in order to design the most appropriate end-effector. Trials using a 500 gram weight (average weight of a Muscat grape cluster) attached to a rachis in place of a harvesting-stage grape cluster and a craft knife have found that a thick rachis can be cut with a force of 100 N.

Friction resistance[1-3.2] was measured by holding a rachis between fingers fitted with sandpaper at a constant force and pulling it in the axial direction until it started to slide. The result shown in Figure 2 indicates that the **friction coefficient**[1-3.2, 3.5] (friction resistance/grasping force) is 1–1.5 and the fingers would not drop a grape cluster weighing 500 grams if a grasping force of 5 N or more was applied.

Fig. 1 Manual grape harvesting

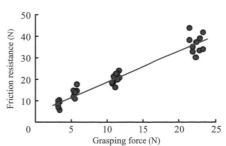

Fig. 2 Relation between grasping force and friction resistance

The **grape harvesting end-effector** prototype unit built on the basis of the above specifications[13, 14] (Movie) and its mechanism are shown in Figures 3 and 4 respectively. This end-effector is comprised mainly of fingers to hold the rachis, cutters, and a pushing device that slides forward and back. It approaches a grape cluster in the horizontal direction to harvest it. The pushing device has three functions: ① it exposes the rachis when the fingers are unable to approach it because it is too short or the cluster has an odd shape, ② it supports the harvesting cluster in transfer to prevent berries dropping and damage from vibrations, and ③ it aligns the cluster in a certain direction.

Figure 5 shows the open-close action of the fingers and cutters. The bed knife of the cutters and one of the fingers are fixed to the end-effector while the opening-closing action of the cutters is linked to the DC motor[(1-3.3)] via gearwheels in the linkage mechanism. The linkage mechanism on the finger side is not fully linked to the DC motor but is constantly subjected to a closing force by a spring. As the cutters open, the pin attached to the cutter linkage mechanism acts on the finger linkage mechanism and the fingers open at the same time. When the cutters close,

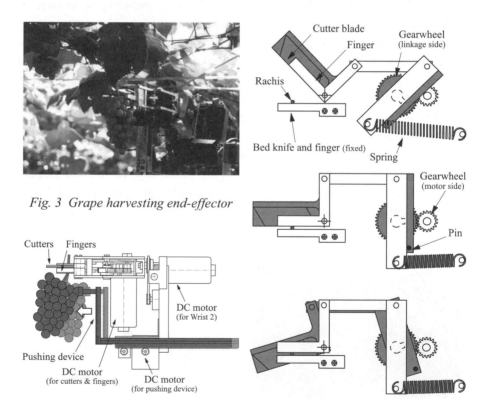

Fig. 3 Grape harvesting end-effector

Fig. 4 Mechanism of harvesting
end-effector

Fig. 5 Action of fingers and cutters

the fingers are closed by the spring. When the fingers close completely and grip the rachis using the spring tension, the pin stops acting on the finger linkage. The cutters close further and sever the rachis. This end-effector is designed to open and close both the fingers and the cutters using a spring and a single motor. The cutting force and the grasping force are set at 100 N and 10 N respectively based on various measurements in relation to the rachis.

Orange and apple harvesting

Much research has been undertaken on the automation of orange and apple harvesting over many years. Well known models include the mandarin orange harvesting robot of Kubota & Co., Ltd. and the orange harvesting robot of the University of Florida, U.S.A., which were announced in 1988 and 1990 respectively. A research report about an apple harvesting robot was published by Cemagref of France in 1985.

The **mandarin orange harvesting robot**[18] was developed and trialed by Kubota & Co., Ltd. This robot, shown in Figure 1, consists of a mobile carriage with a boom, which is fitted with an arm at the end. The arm is a three-degrees-of-freedom **articulated type**[(1-3.6)] having a forearm and an upper arm of equal length. Since the elbow joint and the wrist pitch joint (B and C in Figure 1) are interlinked at a speed ratio of 2:1, the end-effector can perform the translatory movement of a **prismatic joint**.[(1-3.6)] Such a mechanism is reportedly advantageous in that the arm does not retroject like a usual polar coordinate type arm does and therefore does not interfere with trees behind the robot. Figure 2 shows the mechanism and action of the end-effector. First, the strobe light flashes and the camera inside the end-effector acquires an image to detect a fruit. The arm approaches the target

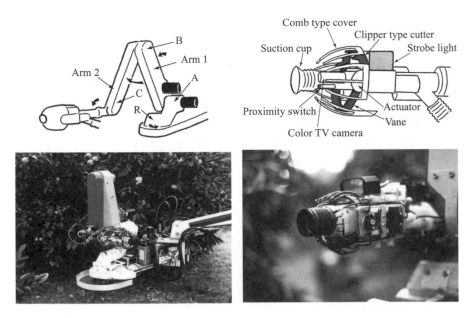

Fig. 1 Mandarin orange harvesting robot[18]

Fig. 2 Mandarin orange harvesting end-effector[18]

Fig. 3 Polar coordinate arm[19] and 7 DOF arm of orange harvesting robots

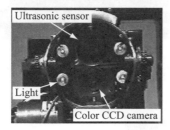
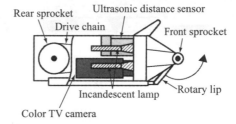

Fig. 4 Orange harvesting end-effector and its mechanism[19]

fruit and grabs it with a **suction cup**.[1-3.5] It takes the target fruit into a comb type cover to separate it from other fruits and cuts its peduncle with a clipper type cutter[1-3.5] (Movie 1).

In the U.S., oranges for processing are efficiently harvested by **shakers,** but the harvested fruits are not suitable for eating due to the impact of vibrations. Hence, research on an **orange harvesting robot**[19] was undertaken at the University of Florida. Figure 3 shows a three-degrees-of-freedom polar coordinate type hydraulic robot, which performs horizontal, vertical and translatory actions, and a seven-degrees-of-freedom arm. Figure 4 shows the end-effector.[1-3.5] The end-effector has a built-in light, a color **TV camera**[1-2.3] and an **ultrasonic sensor**. [1-3.4] It detects a piece of fruit with the camera, approaches in that direction, detects distance with the ultrasonic sensor, and cuts its peduncle by turning the fruit in the direction of the arrow (Figure 4 right) with a semi-circular ring cutter (rotary lip) (Movie 2). The fruits detected by the camera are marked with + on the image in Figure 5. Cemagref in France has also researched an orange harvesting robot[20] (Figure 6), as well as studying an **apple harvesting robot** for many years.[21] Figure 7 shows a prototype robot with two arms[22] which was built by Pellenc s.a. Company based on Cemagref's research.

For these fruit trees, it is necessary to develop cultivation systems which allow robots to detect and harvest the fruits more easily, including dwarfing and **fence shape training**, rather than trying to develop robots that can work on large fruit trees. Such modifications of the cultivation system will make harvesting operation easier for humans as well.

Fig. 5 Image acquired by color TV camera and processed image

Fig. 6 Orange harvesting robot

Fig. 7 Apple harvesting robot[22]

3.9 Strawberry harvesting (annual hill culture)

There are broadly two types of annual hill cultures of strawberries being practiced in Japan: growing strawberries on the ridge wall and growing them on
the ridge top. Research on strawberry harvesting robots began at Utsunomiya
University in 1995 for ridge wall culture and at Okayama University and
Ehime University in 1995 and 1999 respectively for ridge top culture.

The ridge wall cultivation technique involves planting two rows of strawberry plants
on each raised bed so that fruits are set on the slopes of the ridge. The ridge top
cultivation technique involves preparing the plants to set fruits on the plateau of
the ridge. Figure 1 shows a **harvesting robot for ridge wall strawberry culture**
developed at Utsunomiya University. This system uses an end-effector with eight
pneumatically actuated fingers which grasp and detach a fruit from its peduncle by
rotating 270 degrees. To harvest the fruits growing on ridge slopes more easily, a
sheet lifter is fitted to the traveling mechanism to raise the fruits on to a horizontal
plane.[23] In contrast, there is no need to move fruit in ridge top cultivation since
they are already growing on a horizontal plane on top of the ridge. This cultivation
technique is more suitable for mechanical harvesting as there are no obstacles such
as foliage above the fruits.

Both Okayama University and Ehime University used a **Cartesian coordinate
arm**[1-3.6] in their respective studies of a **harvesting robot for ridge top strawberry culture**.[24, 25] This type of arm features a simple structure, high positioning
accuracy and easy coordinate calculation and control. Its capability to approach
the fruit linearly makes it a suitable system for ridge top strawberry culture. The
basic specifications of the arm used at Ehime University are described below as

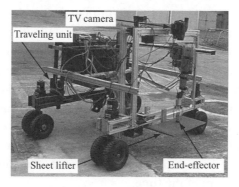

*Fig. 1 Harvesting robot for ridge wall
strawberry culture*

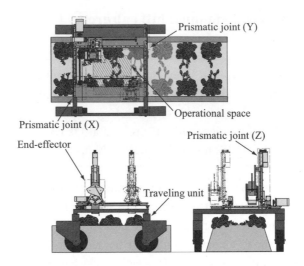

Fig. 2 Harvesting robot for ridge top strawberry culture

an example. The robot is shown in Figure 2. The arm consists of three **prismatic joints**[(I-3.6)] and one **rotational joint**.[(I-3.6)] The length of the axial joint is 520 mm for X, 500 mm for Y and 400 mm for Z. The rotational joint turns 360 degrees in wrist roll motion. The rotation rate and the translation distance of these joints are detected by a **rotary encoder**[(I-3.4)] fitted to each motor. The rotary motion of the motors is transmitted to the prismatic joints and converted to translatory motion via ball screws and trapezoidal screws.

Okayama adopted a **suction type end-effector** and Ehime used a hook type end-effector. Figure 3 shows the suction type end-effector.[24] It consists of a double suction head, an opening-closing cover fitted to the outer cylinder, **DC Motor**[(I-3.3)] 1 to rotate the inner cylinder, DC Motor 2 to drive the opening-closing cover, a **photo-interrupter**,[(I-3.4)] mulch pressers and cutter blades. The upper part of the end-effector has a tube which connects to the suction machine, which creates suction to draw a fruit into the suction head. The adoption of this method has minimized fruit contact and allowed fruit position detection errors made by the **vision sensor** to be compensated. Once the photo-interrupter detects a fruit inside the suction head, DC Motor 2 closes the cover to prevent other fruits from being sucked into the suction head and stop the target fruit from dropping out. DC Motor 1 rotates the inner cylinder fitted with cutter blades to sever the peduncle and harvest the fruit. In the trial, the robot was able to harvest all the targeted mature fruits without damage but it also harvested immature fruits with them approximately 45 percent of the time.

Figure 4 shows the **hook type end-effector**[25] (Movie). It consists of a hook which can pick fruits without using any special cutting device and a fruit receiver

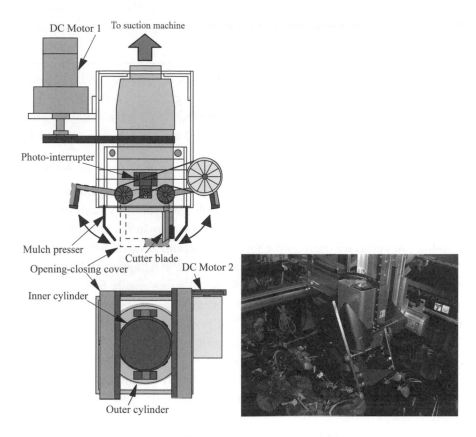

Fig. 3 Suction type end-effector   *Fig. 4 Hook type end-effector*

basket which is interlinked with the vertical movement of the hook in opening and
closing its cover. In the harvesting trial, the fruit position detection error by the
vision sensor was permitted up to 11 mm or so and the incidence of unintended
harvesting of immature fruits was significantly reduced to about 10 percent.

The table top culture system for strawberries is suitable for robotized operation since the fruits are hanging from the culture bed away from the foliage. Research on a strawberry harvesting robot for table top culture began in 1997 at Okayama University and was adopted as one of the projects under the Next-Generation Project from 2003.

The practice of **table top culture**[I-3.3] has been spreading in recent years as it is less onerous for workers than annual hill culture. Figure 1 shows typical methods of table top strawberry culture. The development of the **strawberry harvesting robot for table top culture**[I-3.5] offers the following advantages over other harvesting robots: ① Fruits are set in a smaller space; ② Absence of foliage and other obstacles around the fruits; ③ Longer peduncles; ④ Small fruits enable faster and easier transportation; ⑤ Fruits turn red at the harvesting stage, enabling the use of color imaging.

Accordingly, further expansion of table top strawberry culture will increase the feasibility of the harvesting robot. Among various methods of table top culture, the hanging method shown in Figure 1(b) is more amenable to mechanization and instrumentation than the table method in Figure 1(a) due to the absence of culture bed legs.

Figure 2 shows the prototype robot manufactured at Okayama University[26] which is installed underneath the strawberry bed and uses a three-degrees-of-freedom **Cartesian coordinate type**[I-3.6] arm to perform tasks. The arm moves linearly along the X, Y and Z axes for stroke lengths of 1800 mm, 110 mm and 200 mm respectively. **DC motors**[I-3.3] and reducers were selected for the arm movement speed of 188 mm/s, 295 mm/s and 167 mm/s along these axes, which

Fig. 1 Table top strawberry culture

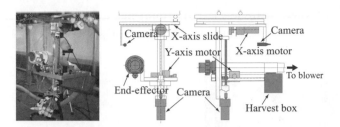

Fig. 2 Cartesian coordinate type harvesting robot and its mechanism

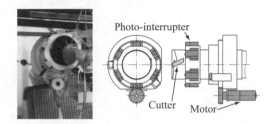

Fig. 3 Suction type end-effector

resulted in a lightweight (6.5 kg excluding the X-axis slide) and compact machine. Fruits mature as they enlarge and hang down on the bed side while immature fruits and flowers grow above mature fruits and on the aisle side. Ideally, harvesting should start from the space below the bed and proceed to the aisle side to minimize obstacles such as immature fruits.

Figure 3 shows the rotary type end-effector. It is a suction type using a vacuum pump so that it can compensate for positioning errors made by the **vision sensor** and its internal diameter is set at 58 mm so that it can harvest fruits of various sizes. Three pairs of **photo-interrupters**[1-3,4] determine whether or not a fruit has been sucked into the suction head, which rotates to position the peduncle in the correct position for cutting. The harvested fruit is drawn into a harvesting tray through a tube by suction created by the vacuum pump (Movie 1). This end-effector can capture a fruit up to 30 mm away from the tip of the suction head as long as the fruit is positioned along the central axis of the suction head.[27] In the experiment, the robot was able to harvest one fruit every 4–7 seconds and harvest all of the targeted fruits. However, it harvested more than one fruit at a time, including immature ones, about 50 percent of the time.

Figure 4 shows the **hook type end-effector**. By fitting the hook at the top of the end-effector to catch the peduncle of the target fruit, this end-effector harvests only the target fruit most of the time (Movie 2).

BRAIN and SI Seiko Co., Ltd. have embarked on the development of the harvesting robot[28] in Figure 5 for commercial application as part of the Next-

Generation Project (Movie 2). This robot has a Cartesian coordinate type manipulator and an end-effector with a finger and a suction plate. It is designed to harvest fruits automatically at night.

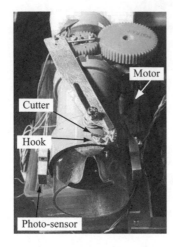

Fig. 4 Hook type end-effector   *Fig. 5 Harvesting robot working at night*

A tomato-harvesting robot has been studied at Okayama University since 1991. It uses a suction cup to pull the target fruit from among several fruits in the cluster and picks it without using scissors. Another type of robot has four flexible fingers.

In Japan, tomatoes for table use are generally grown vertically, supported by poles or strings and set several fruits in a cluster (truss). Figure 1 shows the **tomato harvesting robot**.[29] It has a five-degrees-of-freedom **articulated type arm**[1-3.6] and a two-degrees-of-freedom **prismatic joint**[1-3.6] capable of movement in the forward-backward and up-down directions so that it can approach the target fruit while evading the foliage and efficiently reaching fruit whether set high or low on the plant.

Many cultivars of tomato have a joint called the 'separation (abscission) layer' in its peduncle and the fruit detaches from the plant relatively easily when it is twisted at this joint (Figure 2). In manually harvesting tomatoes, they are either plucked at the separation layer by hand or cut at the calyx with scissors.

Figure 3 shows a **tomato harvesting end-effector**[29(1-3.5)] (Movie 1). It is mainly comprised of a pair of flat bar type fingers, a **suction cup**[1-3.5] connected to a

Fig. 1 Tomato harvesting robot

Fig. 2 Separation layer of tomato

Fig. 3 Tomato harvesting end-effector

vacuum pump, and a **pressure sensor**.[I-3.4] The harvesting procedure is described in Figure 4. It approaches a fruit as it thrusts the suction cup (Figure 4(a)). Once it takes hold of the fruit, it retracts the suction cup in order to pull the target fruit away from other fruits in the truss (Figure 4(b)). Pressure inside the suction cup is constantly monitored by the pressure sensor during this operation. When the pressure reaches the threshold value at which the fruit almost comes off the suction cup, the whole end-effector moves forward at the same speed as the suction cup retraction speed (Figure 4(c)). This way, the fingers move forward while the fruit maintains its position. The fingers grasp the fruit, the wrist joint of the arm rotates to pluck the fruit at the separation layer and the harvesting operation is completed (Figure 4(d)).

Figure 5 shows another type of tomato harvesting end-effector[30, 31 (I-3.5)] (Movie 2). It has a suction cup that moves forward and back and four flexible fingers. As shown in Figure 6, each finger consists of Nylon tubes joined at the top. A cable passes through inside the tubes and its end point is fixed to the tip of the finger. The movement of the cable is synchronized with the forward-backward

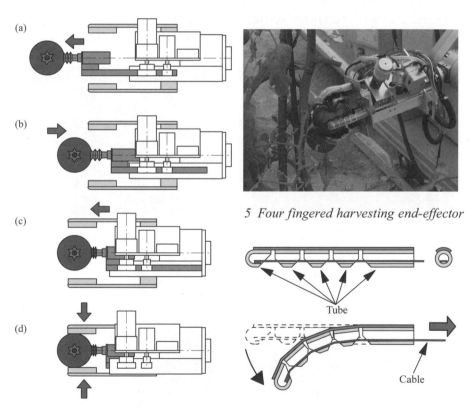

5 Four fingered harvesting end-effector

Tube

Cable

*Fig. 4 Tomato harvesting procedure
by robot*

Fig. 6 Structure and action of finger

movement of the suction cup. As the fruit is drawn into the fingers' operational space, the cable is pulled gradually, folding the fingers along the fruit's surface to grasp it. It is capable of catching the target fruit even when there is little space between it and adjoining fruits or there are many obstacles around it.

> Cherry tomatoes are harvested in the form of cluster attached to the stalk in some cases, but in a majority of cases they are harvested selectively as they mature because fruits ripen progressively from the top to the bottom of the bunch and overripe fruits tend to split. Since they set many fruits and harvesting takes time, Shimane University has conducted research on a robot which selectively harvests red-ripe cherry tomatoes.

Figure 1 shows a **cherry tomato harvesting robot** with a **three-dimensional vision sensor**[I-2.5] (Movie 1). This robot consists of an arm, an end-effector, a 3-D vision sensor and a computer mounted on an electric vehicle which travels on rails laid between ridges. Cherry tomatoes are set side by side on both sides of the peduncle as shown in Figure 2. It is difficult for the end-effector to harvest fruits on the far side of the peduncle because it tends to push the peduncle away as it approaches the fruits. Some clusters may have complex shapes with more than one peduncle and some fruits may be obscured by the foliage. Accordingly, this robot is designed to use a 3-D vision sensor to recognize obstacles to harvesting such as peduncles and foliage as well as mature fruits.

The 3-D vision sensor projects red and near infrared **laser beams** and uses a **PSD**[I-3.4] to receive the light reflected by the target object. It has a pixel count of 120 x 120 and a scanning time of approximately 2 seconds. During scanning, it calculates the distance of each pixel to generate a **3-D image** and compares the photovoltage values of the NIR and red lights to acquire a **binary image**[I-2.4] of

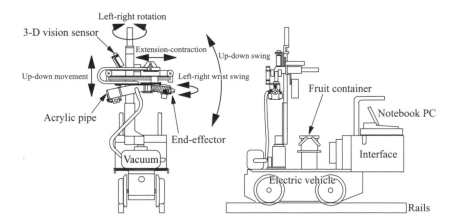

*Fig. 1 Cherry tomato harvesting robot with three-dimensional vision sensor
(Shimane University)*

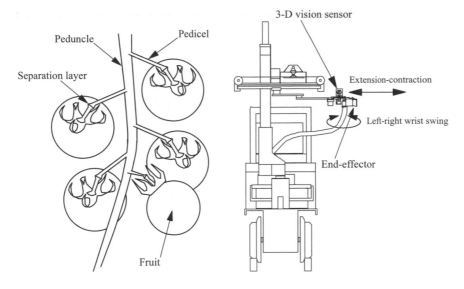

Fig. 2 *Cherry tomato cluster*

Fig. 3 *Cherry tomato harvesting robot with three-dimensional visual sensor on its end-effector (Shimane University)*

red-ripe fruits at the same time. The end-effector draws a fruit to its tip using suction created by the vacuum pump. The diameter of the aperture at the tip of the end-effector is variable. It uses a small aperture to take hold of a fruit, and then opens the aperture wider to pluck the fruit at the separation layer and draw it into an acrylic pipe at the back. When ten harvested fruits are stored in the acrylic pipe, the robot changes its posture to position the pipe on top of a fruit container and opens the pipe cover to drop the fruits into the container. A flexible hose is used so that the end-effector can swing to either side. It processes the 3-D images to recognize foliage and other obstacles and evade them as it harvests fruits. According to its control program, the robot performs scanning by the 3-D vision sensor and harvesting by the end-effector while the vehicle is stationary. Once this harvesting operation is completed, the height of the 3-D vision sensor is adjusted by the arm's vertical movement mechanism to scan and harvest the next fruit and the robot repeats the process.

While this robot scans fruits with the 3-D vision sensor attached to the back of its arm, there are always some fruits that are obscured by the foliage. A study has therefore been conducted on another method by which a smaller 3-D vision sensor is fitted on top of the end-effector as shown in Figure 3 so that it can scan from the side through gaps in the foliage in addition to scanning from the front utilizing the mobility of the arm and the end-effector in order to harvest hidden fruits (Movie 2).[33] First, the arm is retracted to perform frontal scan and the robot recognizes ripe fruits and obstacles and harvest the fruits as it evades the obstacles. When the robot recognizes large leaves in the frontal image which are creating a blind spot, it seeks a gap between the foliage to run another scan from the side and harvests any detected fruits. The test has found a need for some improvement since some fruits were harvested without the calyx.

3.13 Eggplant harvesting (Movie)

> Research to develop an eggplant harvesting robot was undertaken by a group led by the National Institute of Vegetable and Tea Science from 1997 for V-shaped training culture, a cultivation system widely used for eggplant throughout Japan. The robot uses a highly flexible seven-degrees-of-freedom articulated type arm so that it can approach an eggplant from any direction.

In the V-shaped training culture system, the main stem and the first side branch grow at an angle towards the furrows and the fruits hang vertically (Figure 1). It is a suitable cultivation method for mechanical harvesting since more fruits set on the furrow side of the stem and create space between fruits and the stem for fruit grasping and peduncle cutting. Once side branches grow from nodes and become woody, however, they often act as obstacles for the approaching robot arm which are difficult to push away.

For this reason, this **eggplant harvesting robot** is fitted with a seven-degrees-of-freedom arm, which is equivalent to the human arm, and uses **articulated type**[I-3.6] joints for flexibility and large operational space (Movie).[34] The articulated arm has **rotational joints** at the waist, shoulder, elbow, wrist pitch and wrist roll joints as well as roll joints for the upper and fore arms. It is capable of approaching a fruit from any direction. The system is 2200 mm long, 610 mm wide and 2150 mm high and has a controller, a sensing unit, an arm, an end-effector and a traveling vehicle (Figure 2).

A compact **CCD camera**[I-2.1] and an **ultrasonic sensor**[I-3.4] make up the sensing unit, which is fitted to the end-effector. The end-effector shown in Figure 3 consists of scissors and fingers attached to an electric-powered opening and closing hand. When the hand closes, the fingers close first to grasp the peduncle

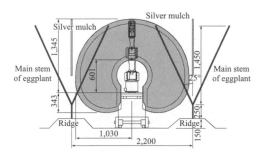

*Fig. 1 V-shaped training system
for eggplant*

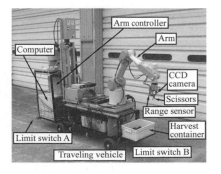

Fig. 2 Eggplant harvesting robot

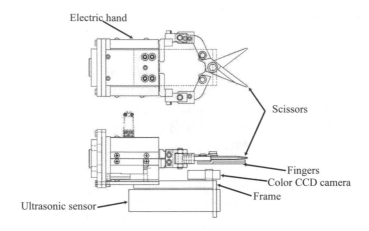

Fig. 3 Eggplant harvesting end-effector

and the scissors close next to cut it.[35] The traveling vehicle can be operated via a computer mounted on the electric work trolley with the **rotary encoder**[I-3.4] and the **limit switch**[I-3.4] which controls forward and backward movement, detects traveling distance and executes emergency stopping.

The following harvesting procedure is employed. The robot performs **global sensing** in the initial position, moves the tip of the arm horizontally and vertically towards the estimated direction of the target fruit and moves 100 mm forward to the fruit. It then performs **local sensing** and controls the tip of the arm so that the gravity center of the target fruit comes to the center of the image and the distance to the fruit is between 160 and 250 mm.[II-2.9] It performs local sensing again in this position to calculate the length of the fruit. If the fruit is at least 120 mm long, it adjusts the arm position so that the stalk end of the fruit[II-2.9] comes to the center of the image and measures the distance to the fruit again using the ultrasonic sensor. It moves the color CCD camera closer to the stalk end of the

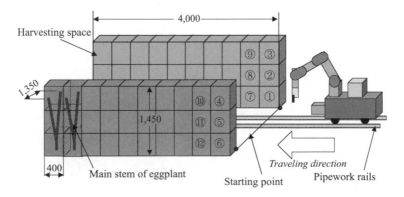

Fig. 4 Harvesting procedure by robot

fruit and inserts its peduncle between the scissors and the fingers located 20 mm above the optic axis of the camera. It closes the electric hand to grasp the peduncle with the fingers and cut it with the scissors. As shown in Figure 4, it repeats this cycle of harvesting, image scanning and processing within the harvesting space divided into three sections vertically and by the crop space along the furrow.

> Cucumber crops are labor intensive because the fruits grow very fast and need to be harvested daily. Joint research on a cucumber harvesting robot was started by Okayama University and ISEKI & Co., Ltd. in 1991. IMAG-DLO of Holland has conducted research on a robot which defoliates as well as harvests.

It is frequently necessary to modify the target crop and its cultivation environment to facilitate the early introduction and efficient operation of agricultural robots. That is, typically the most effective way to create conditions that are suitable for agri-robots, especially fruit harvesting robots, is to change the cultivation system. Existing cultivation systems have developed gradually over many years for the purpose of improving productivity and quality. Improving workability has not been very important perhaps because humans with high intelligence and adaptability have been able to work efficiently enough in the existing cultivation systems. It is difficult to expect a marked improvement in the function and performance of agricultural robots over the current level due to cost factors. If we are to derive a level of output efficiency equal to or better than the human level from the robots, we need to consider possible changes required for a transition from crop-centered cultivation systems to more robot-friendly ones in order to create an appropriate manner of interface between robots and crops.

There are currently two common cultivation methods of cucumbers: the vine hanging system and the top pinching system. The vine hanging system is mainly used to prolong the harvesting period during the cool season and involves clipping two or three vines growing from each plant to a string stretched high above the bed, an operation which is repeated as the vines grow longer. The top pinching system, in contrast, is mainly used for a high yield within a short period of time during the warm season and involves the use of a cucumber net. The top of the main stem of a cucumber plant which grows above the cucumber net is pinched out and the secondary vines are allowed to grow up to two or so nodules before the growing tip is pinched out.

In these cultivation systems, cucumbers grow vertically. When they are hidden behind large leaves, even humans sometimes overlook the fruits. It is more difficult for the vision system of a robot to detect fruits that are partially or totally hidden by the foliage because its capability to discriminate similar-colored objects or determine and understand forms is less than that of humans. Robotic arms also generally have fewer degrees of freedom than the human arm and are often hindered by the foliage when approaching fruits. When the peduncle, stem

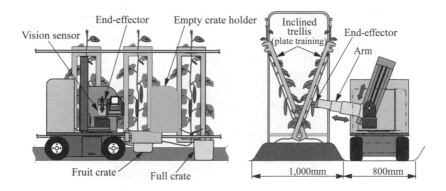

Fig. 1 Inclined trellis training system and cucumber harvesting robot
(Okayama University, ISEKI & Co., Ltd.)

and petiole of the cucumber plant are located close together, the end-effector may accidentally cut them together when harvesting.

For these reasons, a cultivation system for easy separation of fruits and foliage is being researched so that the control and mechanism of the robot can be simplified and the robot can harvest more easily.[36] Figure 1 depicts this cultivation system and a **cucumber harvesting robot**. This is called the **inclined trellis training system**[1-3.2] by which the conventional cultivation system is tilted and the foliage is held up by support posts so that only fruits hang below the trellis. This system is considered suitable for harvesting by the robot because the fruits and the foliage are separated.

Cucumber harvesting robots are being studied from the perspectives of both agronomics and engineering in efforts to construct a harvesting system for this inclined trellis training system. Figure 2 shows this cucumber harvesting robot in operation (Movie 1). It consists of a **vision sensor** to discriminate, recognize and detect the position of the fruit, a seven-degrees-of-freedom **polar coordinate type arm**,[1-3.6] an end-effector[1-3.6] to grasp the fruit, detect and cut the peduncle, and a traveling mechanism.[36]

The arm has a one-degree-of-freedom **rotational joint**[1-3.6] (base waist), an up-down slide which translates parallel to the inclined trellis, one **prismatic joint**[1-3.6] (slide) and four rotational joints (waist, shoulder, wrist pitch and wrist roll) in order to harvest cucumbers growing on the inclined trellis on either side of the furrow. It uses a prismatic joint in the direction of approaching the fruit because it can approach the fruit in a linear fashion due to the separation of the fruit and the foliage and the absence of obstacles such as foliage in the vicinity of the fruit under this system of cultivation. The sliding joint is comprised of three sections: the upper arm on the shoulder side, the middle arm, and the fore

Fig. 2 Cucumber harvesting operation by robot

arm on the tip side. The rotational motion of the AC servo motor fitted to the upper arm is converted to the linear motion of the middle arm via the ball screw and guide rail. The middle arm has two sets of timing pulleys and belts in the direction of its sliding motion; one is fixed to the upper arm and the other to the fore arm. This way, the linear motion of the ball screw is transmitted to the fore arm, which can extend and contract at twice the speed of the middle arm. This mechanism makes it possible to reduce the distance between the shoulder joint and the wrist pitch joint to 300 mm.

Figures 3 and 4 show the harvesting end-effector and its mechanism respectively. The end-effector consists of two flat bar fingers to grasp the fruit near its upper end, a **photo-interrupter**[1-3.4] to detect the fruit in the grasp of the fingers, and a peduncle cutter to detect and cut the peduncle by sliding forward-backward and up-down over the surface of the fruit. The peduncle cutter has a U-shaped guide with a 10 mm wide notch and a cutter blade behind it. The U-shaped guide can move forward and back powered by a **DC motor**[1-3.3] and a trapezoidal screw thread and up and down by parallel linkage. After the photo-interrupter detects the position of the fruit, the end-effector grasps the top part of the fruit and moves the peduncle cutter forward. When the tip of the U-shaped guide contacts the fruit, it slides upward along the fruit since a fruit more than 20 mm in diameter does not fit into the guide. When the guide reaches the part of the fruit that is narrower than the width of the notch, i.e., the peduncle, the guide takes it into the notch. The cutter blade, which is actuated by a spring force, contacts the peduncle and cuts it.

As above, the peduncle cutting mechanism consists of a series of mechanical procedures based on the **physical properties**[1-3.2] of the cucumber. They ensure that harvesting operation is executed consistently without being affected by the plant's ambient temperature, humidity, scraps or dust. The harvesting experiment achieved a 91 percent success rate for harvesting action alone and a harvesting rate

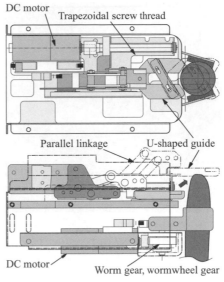

Fig. 3 Harvesting end-effector

Fig. 4 Mechanism of harvesting
end-effector

of 77 percent for the sequence of harvesting operation, including fruit recognition by the vision sensor, under the inclined trellis training system.

In Holland, which has highly-advanced horticultural technology, IMAG-DLO has conducted research on a cucumber harvesting robot, which is shown in Figure 5 (Movie 2).[37] It has a 7-DOF **articulated type**[(I-3.6)] arm, which slides in parallel to the traveling direction on the traveling mechanism so that it does not interfere with the field of the vision sensor. Since the most common cultivar of cucumber in Holland has a long peduncle over 50 mm in length, the end-effector grasps the peduncle and cuts it with electric heat cutters.[(I-3.5)]

Research is being conducted on the use of this robot with a different end-effector for defoliation (Movie 3). In the defoliation operation, older leaves with a reduced photosynthetic ability in the lower part of the main stem are removed. One to two leaves per stem are usually removed in a single operation. This system also defoliates around the target fruits so that they are more exposed for efficient harvesting operation.

This end-effector consists of three electric heat cutters and an air actuator which is positioned to surround the main stem. First, the vision sensor recognizes the leaf to be removed and the stem it is attached to and detects their positions. The end-effector approaches the stem below the leaf as shown in Figure 6 and positions it so that the electric heat cutters surround the stem on three sides.

Fig. 5 Cucumber harvesting robot

Fig. 6 Defoliation procedure
(arm approach to stem)

There is a V-shaped guide above the cutters which slides upward along the stem and inserts the petiole of the leaf into one of the three cutters. When the petiole contacts the cutter, electricity is turned on and the cutter burns through the petiole to remove the leaf.[1-3.5] Figure 8 shows the petiole cutting procedure. Since the cut section is sterilized with heat at the time of cutting, this method can prevent the plant from contracting disease through the cut section.

Fig. 7 Defoliation procedure (electric
heat cutter position)

Fig. 8 Defoliation procedure (cutting
by electric heat cutter)

Since lettuces do not form heads at the same time and some heads grow large but are only loosely formed inside, growers tend to selectively harvest them by checking the firmness of individual heads with their fingers. The lettuce harvesting robot developed at Shimane University has fingers with a force sensor as well as a three-dimensional vision sensor to detect the position and size of the lettuce so that it can selectively harvest large and tightly formed lettuces.

Figure 1 shows a robot for selectively harvesting head lettuce (Movie).[38] It consists of a **Cartesian coordinate arm**,[1-3.6] a harvesting end-effector, a **3-D vision sensor**[1-2.5] for lettuce recognition and a computer, all mounted on a four-wheel-drive electric vehicle which travels straddling the ridge. The 3-D vision sensor scans the ridge from above with a **laser beam** from a **laser displacement meter** as shown in Figure 2. The laser displacement meter measures distance by **triangulation**[1-2.4] using a near infrared laser beam (at 830 nm wavelength) and a **PSD**.[1-3.4] Combining a scan perpendicular to the ridge by a rotating polygon mirror (hexagonal mirror) and one parallel to the ridge by the arm movement along the x-axis, the unit generates a **three-dimensional image** of 100 x 100 pixels.

Figure 3 is a flow chart of the harvesting operation. The end-effector (Figure 1 right) cuts the stem and harvests the lettuce. The bottom of the head lettuce is generally 2–4 cm above the ground and cutting any higher than that would cut into the head itself. Accordingly, the end-effector descends to the side of the lettuce and uses two V-shaped lower fingers to hold down the bottom of outer layer leaves so that the cutter blade is positioned at a height of approximately 2 cm above the

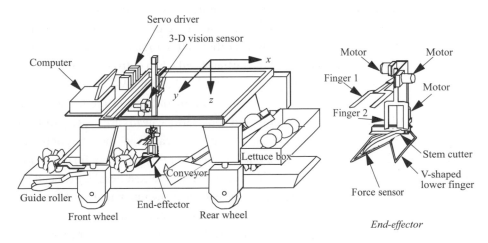

Fig. 1 Lettuce harvesting robot

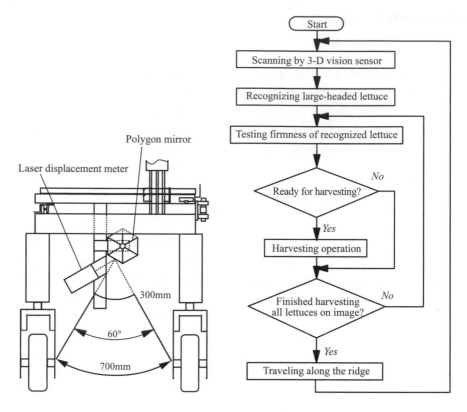

Fig. 2 Three-dimensional vision sensor   Fig. 3 Flowchart of lettuce harvesting
                                                            operation

ground. It stops descending when the ground reaction force exceeds a certain value. Next, the end-effector approaches the lettuce to position the stem between the two V-shaped lower fingers and cuts it. The stem is cut by the movement of the cutter blade powered by a motor but the cutting force occasionally pulls out the roots depending on the soil type or the condition of the roots. To prevent this, the end-effector uses its fingers to grasp the lettuce head when cutting the stem. After stem cutting, the head is carried by Finger 1, Finger 2 and the V-shaped fingers to a conveyor belt. The system recognizes the position and size of the lettuce head using its 3-D image processing program but some lettuce heads are loosely formed even though they have large diameters or they may appear larger than they actually are when they are covered with large outer leaves. For this reason, a **force sensor** is fitted to one of the V-shaped lower fingers to check the firmness of the head by pressing it lightly so that the robot only harvests those heads that are sufficiently large and firm based on the 3-D image and the force sensor result.

The robot takes 2 seconds for 3-D scanning, 0.9 seconds for image processing, 1 second for firmness testing by the force sensor, 1.5 seconds for harvesting by the end-effector and 1 second to carry the harvested lettuce to the conveyor. The robot vehicle stops for recognition and harvesting and travels along the ridge to move on to the next lettuce. In the field experiment, it took 11.3 seconds on average to harvest one lettuce, including the traveling time and the time for skipping immature heads. It harvested 94 percent of the harvesting-stage lettuce heads. The remaining lettuces were mostly covered by outer leaves and the 3-D vision sensor was unable to measure the size on the images. The experiment was conducted using a commercial power source of AC 100 V but the robot needs to carry a small generator in the actual harvesting operation in the field.

3.16 Cabbage harvesting

> While selective harvesting of head-forming large-mass vegetables such as cabbage is hard physical labor, the quality standards demanded by the market are becoming more stringent. Robotization of this operation requires a system capable of determining the harvesting stage and picking cabbages without affecting their quality. NARC has been studying a robot for selectively harvesting cabbages since 1994.

Cabbages generally grow at variable rates and there are considerable size differences between individual cabbage heads in the harvesting season. Accordingly, selective harvesting is practiced in many cabbage producing regions. Also, head-forming vegetables such as cabbage tend to be large and heavy and the harvesting, processing and boxing operations are usually performed at the farm. It is extremely hard labor for aging farm workers who must labor in difficult postures and repeatedly carry their harvest off the farm. For these reasons, research has been conducted on a **cabbage harvesting robot** that recognizes harvesting-stage cabbages by imaging[II-2.11] and selectively harvests them (Movie). Its arm and end-effector have some special features for efficient harvesting and transportation since this robot has to handle a head-forming vegetable weighing over 1 kg (in contrast, fruit harvesting robots handle crops weighing only 100 g or so).

Figure 1 shows the cabbage harvesting robot developed by NARC.[39, 40] A hydraulic arm fitted with a harvesting end-effector is mounted on a traveling mechanism based on a crawler type carrier vehicle with a 1100 mm tread.

Figure 2 shows the arm's mechanism. The work speed and light weight of the arm, although it has to handle a heavy object, are important for efficient harvesting by the robot. The arm is a **polar coordinate type**[I-3.6] having a total

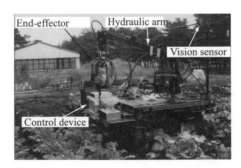

Fig. 1 Cabbage harvesting robot

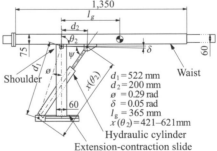

$d_1 = 522$ mm
$d_2 = 200$ mm
$ø = 0.29$ rad
$δ = 0.05$ rad
$l_g = 365$ mm
$x(θ_2) = 421 – 621$ mm
Hydraulic cylinder
Extension-contraction slide

Fig. 2 Structure of hydraulic arm

of four-degrees-of-freedom, consisting of two-degrees-of-freedom of **rotational joints**[1-3.6] at the waist and the shoulder, one-degree-of-freedom of a **prismatic joint**[1-3.6] which extends and contracts the arm, and one-degree-of-freedom of a free joint at the end-effector mounting section on the tip without an actuator. Compared with the articulated type arm,[1-3.6] the polar coordinate type arm can have moving parts with relatively low positioning accuracy because positional control errors of the individual moving parts do not accumulate at the end. Also, the arm is designed to allow some degrees of deflection under load at the end to keep its weight down. The test-manufactured arm weighs a total of 16.4 kg, which is relatively light-weight for a heavy-lifting device, and the arm deflection was 4.1 mm at a static load of 196 N applied by the end-effector and the cabbage at 2100 mm from the shoulder joint. **Hydraulic motors**[1-3.3] are used as actuators for the waist and the extension-contraction slide and a **hydraulic cylinder**[1-3.3] linkage is used for the shoulder which generates large torque. The rotational motion of the hydraulic motor is converted into the linear motion of the slide by a ball screw.

Figure 3 shows the mechanism of the harvesting end-effector. It can operate two head grasping fingers and two stem cutting fingers independently. The grasping fingers clutch the cabbage head, and then the cutting fingers sever the stem. The opening between the two grasping fingers is approximately 300 mm, which is sufficient for the harvesting of mature cabbages around 200 mm in diameter. Both sets of fingers are actuated by spring-return type single-action hydraulic cylinders, which generate a thrust that is converted to rotational motion by a linkage mechanism to open the fingers. The use of the spring-return type cylinders means that only one pipe is required and that the hydraulic circuit can be

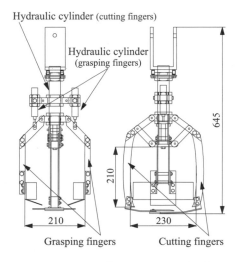

Fig. 3 Structure of harvesting end-effector

simplified to keep the end-effector light and compact. The cabbage grasping force and the stem cutting force are set at approximately 92 N and 532 N respectively through regulation of the cylinder relief pressure.

In order to harvest the cabbage without affecting its quality, the robot must grasp the head in an appropriate position when cutting its stem. It needs to acquire information about the distance between the end-effector and the cabbage head and the field surface but information from the **vision sensor** alone is insufficient. It is therefore fitted with an **ultrasonic sensor**[II-3.4] and a ground contact switch at the base of the end-effector to measure the distance to the head and a contact sensor to detect the grasping condition.

The harvesting procedure is as follows. The system determines the harvesting-readiness of the cabbage and estimates its 2-D position from image data. Then it moves the end-effector to a position directly above the cabbage and lowers it. If the finger tip touches the ground or the ultrasonic sensor continues to recognize the target object for over a set duration (0.9 seconds) after it detects it at a distance of 200 mm, the system recognizes that the cabbage is within the grippers' range and harvestable. It grasps the cabbage, raises it by 100 mm to cut the stem with the cutting fingers, and harvests it.

Watermelon is both heavy and fragile, and therefore requires careful handling. Research on a harvesting robot which handles heavy fruits and bagged fertilizers began at Kyoto University in 1994. Research on the tele-operational robot for watermelon harvesting began at Sung Kyun Kwan University, Korea.

The watermelon bears heavy fruits, which need to be transported over a relatively long distance on the farm. Research on a watermelon harvesting robot has been conducted at Kyoto University for the purpose of labor-saving.[41–43] Figure 1 shows a **watermelon harvesting robot** (Movie). This robot mainly consists of an arm with a linkage mechanism and four joints as shown in Figure 2, an end-effector, a base eye and a traveling unit. At the first joint of the arm, which requires high rigidity and high **resolution**, the revolution of the hydraulic radial piston motor is transmitted to the harmonic drive reducer via timing belts and pulleys. At the second joint, which requires high output, the motion of two **hydraulic cylinders**[(I-3.3)] is transmitted to the crank. At the third joint (prismatic), which requires high resolution and high speed, the motion of a **DC motor**[(I-3.3)] with a reducer is transmitted via a timing pulley and belt. At the fourth joint, the third joint's slider is fitted with a bearing to serve as a passive joint in order to minimize the **friction coefficient**[(I-3.2)] of the joint. This way, the arm and the end-effector are equipped with mechanisms and structures based on a control design aiming for a lower cost.[44]

More specifically, applying **PID control**[(I-4.4)] to each shaft of this arm and the end-effector would not achieve the required performance. A success rate of 86.7 percent and an operation time (including loading time) of 14 seconds per fruit can be achieved by the application of robust control theory.[42] Figure 3 shows the procedure to load a watermelon on the tray of the **transport vehicle**. The wires are

Fig. 1 Watermelon harvesting robot

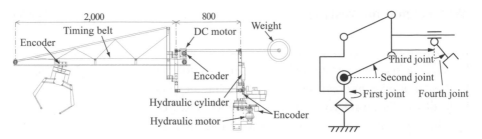

Fig. 2 *Mechanism of arm²*

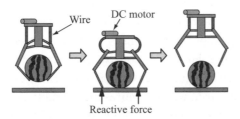

Fig. 3 *Action of end-effector⁴²*

tight when the end-effector is holding the fruit as in the left diagram of Figure 3. With the action of the second joint, the fingers contact the tray and the reactive force on the finger tips releases the fruit. The loose wires are rewound by the light-weight and low-output DC motor and the end-effector is raised.

At Sung Kyun Kwan University in Korea, a **tele-operation** robot[II-4.24] using human vision is being tested for watermelon harvesting.[45–47] Figure 4 shows the system consisting of a four-degrees-of-freedom **Cartesian coordinate arm**[I-3.6] attached to a **gantry**[II-4.19] (1600 mm x 1600 mm x 3300 mm, max speed 1 m/s).

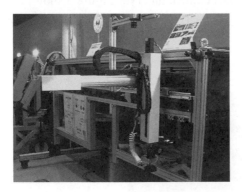

Fig. 4 *Tele-operational robot with gantry system*

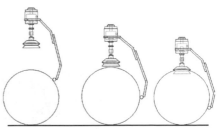

Fig. 5 *Action of watermelon harvesting end-effector*

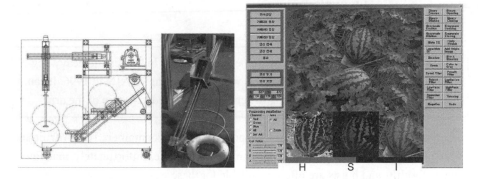

Fig. 6 Fruit loading device   *Fig. 7 Images used for tele-operation*

Figure 5 shows the action of the end-effector, which uses a **suction cup**[1-3.5] (diameter 120 mm, vertical stroke 22 mm, ball joint-based cup angular variability 30°) to change the orientation of the fruit. Figures 6 and 7 show the fruit loading device and the images used for tele-operation respectively.

3.18 Heavy material handling

A device to replace human labor with mechanical labor in harvesting heavy vegetables and transporting farm materials was developed at BRAIN in 1996. It can perform various tasks by replacing the end-effector fitted to the tip of the arm.

In harvesting heavy vegetables and transporting/loading containers full of fruit and vegetables, many tasks are too heavy for human power alone. Unfortunately, machinery such as forklifts and hoists are often unsuitable for these tasks due to space limitations or lack of positioning accuracy.

BRAIN therefore developed an **operation assisting device** which uses a powerful arm fitted with an end-effector to perform medium-heavy material handling with an operational accuracy that is enhanced by the human operator who operates the end-effector[48] (Movie).

The basic mechanism of the arm is a **polar coordinate type**,[(1-3.6)] which is considered to be better in terms of its weight, range of motion and safety and has three-degrees-of-freedom (vertical rotation axis, horizontal rotation axis and telescopic axis). Each axis has a servo motor as its actuator, which provides excellent control capability. It controls speed in proportion to the tilt angle of the joystick located at the tip of the telescopic arm to perform pushing and pulling as well as lifting. Further, the arm is fitted with a tilt sensor and a displacement sensor to perform simultaneous multi-axial control by a programmable controller in order to facilitate vertical and horizontal motions.

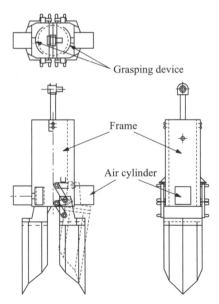

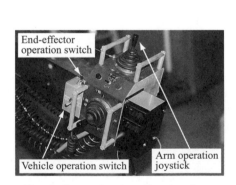

Fig. 1 Control unit at the tip of arm

Fig. 2 Daikon radish harvesting end-effector

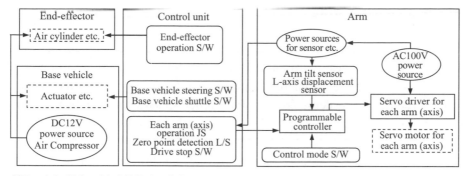

S/W: switch, JS: joystick, L/S: limit switch.

Fig. 3 System configuration of assisting device

Various end-effectors can be available for attaching to the arm via a simple joint, including a pneumatic ***daikon*** radish **harvesting end-effector** (Figure 2), a **cardboard box handling end-effector** and an **watermelon handling end-effector**. When a **tractor** capable of automatic steering and forward-backward travel/stop switching (shuttle) is used for mounting and transporting the arm, it is possible to control the vehicle remotely from the arm control unit for better mobility and operability in the field.

It has been confirmed that the application of the farming assist device configuration in Figure 3 to *daikon* radish harvesting (Figure 4) can result in a reduction in labor burden compared to the conventional work procedure (human labor) without causing any significant damage to the harvested crops. One assist device for medium-heavy handling commonly used at fruit grading facilities is Raku-Raku Hand (Aikoku Alpha Corporation)[49] (Figure 5). It is an electric powered device fitted with a hook at the tip of the arm to help workers lift and carry a container with less force. There is a type of heavy material handling device called a 'balancer', which can be fitted to a light 4WD truck tray (e.g., Balaman for fruit handling by Toyo Koken Co., Ltd.[50]) (electric, pneumatic etc.).

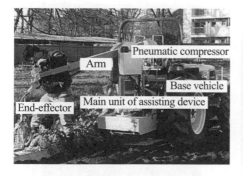

*Fig. 4 Daikon radish harvesting with
assisting device*

*Fig. 5 Raku-Raku Hand for
container handling*

> A mushroom harvesting robot, developed by Silsoe Technology Limited, U.K., is close to the commercialization stage. It selects a harvesting-stage mushroom from the dense growth of mushrooms, takes hold of it by the cap using the suction cup and twists it out of the culture medium.

Button mushrooms are the most widely cultivated mushrooms for table use in the world. Global production is estimated at 1.4 million tons. The mushrooms are grown on shelves (Figure 1), boxes or bags filled with compost made from stable manure and rice straw and repeatedly appear at seven to ten day intervals (in 'flushes'). As they grow fast, any delay in harvesting will allow them to grow too large and reduce their commercial value. In manual harvesting, the worker lightly holds the cap of a mushroom with fingers and twists it out of the compost. When the mushrooms are very dense, the worker must hold down other mushrooms with the other hand gently so not to lift them out of the compost.

Figure 2 shows the **mushroom harvesting robot**[51(I-3.5)] developed by Silsoe Technology Limited, U.K., which is comprised of a vision unit, a three-degrees-of-freedom **Cartesian coordinate arm**,[(I-3.6)] an end-effector and a conveyor (Movie). Since mushrooms grow on an almost two-dimensional plane and there are virtually no obstacles other than other mushrooms in the vicinity, the 3DOF Cartesian coordinate arm can perform necessary tasks adequately. The vision unit uses a monochrome camera to recognize mushrooms on an acquired image (Figure 3) and determines the optimum order of harvesting based on adjacency conditions and space availability.

Figure 4 shows the harvesting end-effector. When there are adjoining mushrooms, it is very difficult for fingers to grasp a mushroom without damaging

Fig. 1 Mushrooms grown on shelves　　　*Fig. 2 Mushroom harvesting robot*

Fig. 3 Image acquired by machine vision   Fig. 4 Harvesting end-effector

the cap. For this reason, this end-effector lowers a **suction cup**[1-3.5] using a rack and pinion mechanism down to the target mushroom (stroke 250 mm), takes hold of the cap, rotates 180 degrees and tilts the cap and the stem together to pull it out of the compost. Large and small suction cups are attached to a rotary mechanism and are interchangeable according to the size of mushroom cap. The mushroom contact area of the suction cup has cushioning material so that it does not leave a mark on the cap.

Harvested mushrooms are carried to the conveyor in Figure 5. This device has a circulating chain with fingers attached at regular intervals. The polyurethane fingers can grip a mushroom stem approximately 20 mm to 70 mm in diameter. It has a high speed rotary cutter which cuts the stems of passing mushrooms below the fingers. At the end of the line, a pair of grippers, which is capable of open-shut and forward-back movements, holds a mushroom at the cap and passes it on to the next conveyor or places it in an appropriate container according to its size (Figure 6).

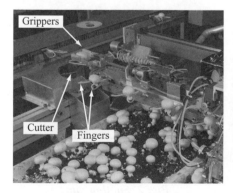

Fig. 5 Conveyor for harvested   Fig. 6 Grippers
           mushrooms

3.20 Fruit grading

SI Seiko Co., Ltd. released a commercial robot system in July 2002 which automatically removes near-spherical fruit such as peaches, *nashi* pears and tomatoes from their containers and shifts them to a conveyor. Nishijima Co., Ltd. has released a commercial fruit packing robot that can pick up fruits from the conveyor tray.

In the conventional practice of grading deciduous fruit such as peaches, *nashi* pears and apples or tomatoes, a worker manually picked fruit from its container, looked at the rear side of the fruit to determine its grades and roughly sorted it as they placed it on the conveyor trays by way of tracking. Recently, this process has been replaced by a fruit providing robot which automatically feeds fruits in flat containers to the middle stage and a **fruit grading robot** which transports them from the middle stage to the conveyor trays[52(II-2.14–16)] (Figure 1).

The fruit providing robot uses a three-degrees-of-freedom **Cartesian coordinate arm**[(I-3.6)] with prismatic joints[(I-3.6)] in three directions (X, Y and Z) developed by SI Seiko Co., Ltd. and an end-effector with six **suction cups**[(I-3.5)] as shown in Figure 2. The robot arm descends (Z-axis) to suction five or six pieces of fruit at a time, then ascends and shifts the fruits horizontally (Y-axis) to match the spacing between conveyor trays. Then, it moves horizontally (X-axis) to the middle stage and descends to release the fruit to the middle stage (Figure 3). The stroke of this robot is 900 mm and each reciprocating motion takes four seconds. In this system, two robots handle containers independently of each other to provide twelve pieces of fruit to the middle stage. One fruit grading robot in the next stage handles these twelve pieces of fruit at once (Movie 1).

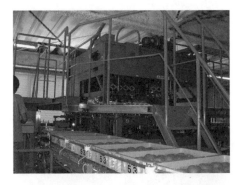

Fig. 1 Fruit grading robot, fruit providing robot and fruit containers (SI Seiko Co., Ltd.)

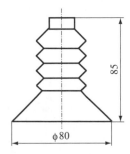

Fig. 2 Suction cup

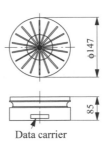

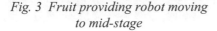

Data carrier

Fig. 3 *Fruit providing robot moving*   Fig. 4 *Fruit conveying tray*
        to mid-stage

The fruit grading robot uses a 3DOF Cartesian coordinate arm equipped with X-axis and Z-axis prismatic joints and a **rotational joint**[1-3.6] on the end-effector. The end-effector has twelve suction cups. The robot descends (Z-axis) and suctions twelve pieces of fruit at once, then travels (X-axis) over a light and a **TV** camera.[53(II-2.3)] Since the spacing between fruits is the same as the spacing between trays waiting on the conveyor, the fruit grading robot does not need freedom along the Y-axis. It stops the movement of the fruit at a distance of 290 mm above the waiting conveyor trays and rotates 270 degrees around the center axis of each suction cup. During this rotation process, twelve cameras take lateral view images of the twelve pieces of fruit at 90-degree intervals simultaneously. The fruit is released onto the conveyor trays shown in Figure 4. The base of each conveyor tray is fitted with a readable **data carrier** (memory volume 256 bytes) that can record the **grade and class** information about each piece of fruit. The conveyor trays are carried on a free-flow type conveyor belt traveling at a speed of 30 m/min (Movie 2).

The stroke length of this arm is 1165 mm and the maximum speed is 1000 mm/s. Speed is controlled as shown in Figure 5 between the fruit suction point to the tray release point (A to E) and takes 2.7 seconds to complete. Additionally, it takes 0.4 seconds to descend from the default position to the fruits (B to A), 1.0 second to return to the default position from the tray release position (E to B) and 0.15 seconds for standby time. The total time is thus 4.25 seconds per cycle. This system can process about 10,000 pieces of fruit per hour. The vacuum pump used to create suction has a rated output of 1.4 kW, a rotating speed of 3400 rpm, a vacuum pressure of up to 3800 mmAq and a displacement of 1.3 m^3/min. One pump supplies suction to six suction cups. By setting the suction pressure at 30 kPa for peaches and 45 kPa for *nashi* pears and apples, it is capable of handling peaches even more gently than inexperienced human fingers.

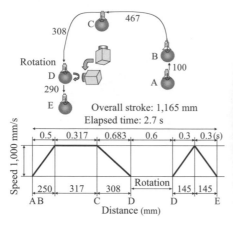

Fig. 5 Stroke and speed of fruit
grading robot

Fig. 6 Tomato fruit boxing robot
(Nishijima Co., Ltd.)

A new type of tomato **boxing robot** is in practical use. The old type of robot picked tomatoes from stationary trays by suction and placed them in boxes but the new model can pick fruits from moving trays for packing as shown in Figure 6 (Movie 3).

Agricultural products are often sold in a tied bundle of several pieces or in a bag of a certain quantity in the shop. This section describes the full-automatic filling and binding machine for grains and the asparagus binding machine commercialized by Omi Weighing Machine Inc.

Agricultural products are sold in a wide range of packages, including bags, trays, boxes and bundles, depending on the product type and the distribution process. Since packaging is time and labor intensive, various methods of mechanization are being introduced. This section describes a **fully-automatic filling and binding machine** for grains, a **binding machine for asparagus** and the mechanism used for the binder.

Figure 1 shows the fully-automatic filling and binding machine manufactured by Omi Weighing Machine Inc.[54] which automatically weighs, fills and binds paper bags for grains such as rice, wheat and soy. First, a bag (up to 100 bags can be set in the holder) is picked up by two **suction cups**[1-3.5] and carried to the grain filler. Two fingers receive and open the bag. The filler spout descends and fills the bag with grains while shaking the bag up and down for even filling (Figure 2). Once a designated quantity has been filled, the bag is carried on a conveyor to the folding stage. A pair of long flat bar type fingers hold the bag opening from both sides and rotate downward to fold it (Figure 3).

The machine presses the top of the bag down to form shoulders, pulls the fingers out of the fold and moves the bag to the binding stage on the conveyor. The tie strings attached to either shoulder are twisted 90 degrees (Figure 4(a)). Arms A fold the tips of the strings (Figure 4(b)). Arms B grasp the tips and draw them

Fig. 1 Full-automatic filling and binding machine for grains

Fig. 2 Filling procedure

Fig. 3 Bag opening folding procedure

outwardly (Figure 4(c)). At this point, the strings have been tied twice. A pair of fingers move forward, Arms B move in an arc while still holding the string tips and wind the strings around the fingers to create loops. Arms B pass the strings to the fingers, each of them holding a part of the other finger's loop through its own loop. The fingers draw the strings to complete a bowknot (Figure 4(e)). The procedure is slightly different from that of bow-knotting by human hands but it produces an identical result. The bound bag is transported by a variable speed belt conveyor and stacked by a palletizing device (Movie 1). The machine handles 30 kg bags and processes 120 bags per hour. Information on the day's work can be checked on the touch panel.

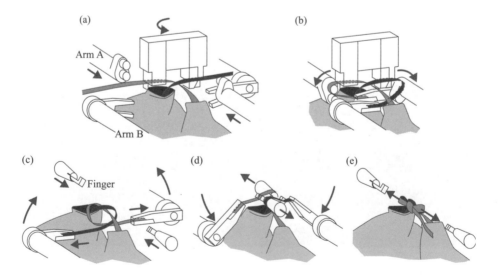

Fig. 4 Drawstrings tying procedure

Fig. 5 Binding machine for asparagus

Figure 5 shows a binding machine for asparagus (Omi Weighing Machine Inc.). A bundle of asparagus spears fed into the machine is transported by an arm and received by fingers. Two discs are fitted in parallel, each with four pairs of fingers positioned at ninety-degree intervals. They turn as they hold a bundle of asparagus at the top and the bottom. When a bundle passes the binding mechanism, both ends are bound with a piece of adhesive tape before being carried away on the conveyor (Movie 2). This machine processes up to 1400 bundles per hour.

The binder uses a unique mechanism called a **knotter bill** (Figure 6). When a bundle of harvested crops reaches a certain quantity, a piece of string is wound around it. The arm (knotter bill) rotates and pulls the string to form a loop. A hook type finger holding part of the string passes through the loop to make a knot. It cuts the string and completes the binding of a crop bundle (Movie 3).

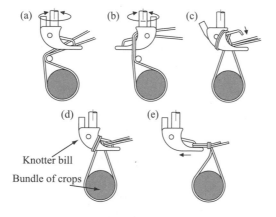

Fig. 6 Binding procedure

3.22 Wool shearing (Movie)

> A wool shearing robot was developed at the University of Western Australia in 1987 with the support of the Australian Wool Corporation. A robotic wool shearing system to shear sheep by two robots was launched by Merino Wool Harvesting Pty Ltd.

Sheep shearing requires the skill of an experienced shearer because the moving sheep must be held down. The operation also requires many workers. University of Western Australia and Australian Wool Corporation began an attempt to robotize this operation to reduce costs in around 1987.[55-57] Figure 1 shows the **wool shearing robot** and the sheep mounting device. Firstly, the operators place a sheep on the mobile platform in a supine position and immobilize its head, forelegs and hind legs with clamps. The platform automatically moves into position below the robot, rotates the sheep and shifts its limbs so that the robot can shear the wool section by section. The head clamp is fitted with a blindfold to cover the sheep's eyes to reduce the sheep's agitation from the movement of the robot. Since every sheep is different in body shape and size, these data are entered in the computer beforehand, then checked and adjusted by **machine vision**.[I-2.1] Figure 2 shows a photo of a sheep and the shearing robot.

Figure 3 shows the hair clipper end-effector. This is called a cutter assembly and is fitted with a drive motor, a follow-up actuator and a sensor to detect the distance between the cutter and the skin to avoid injuries. It is desirable for the

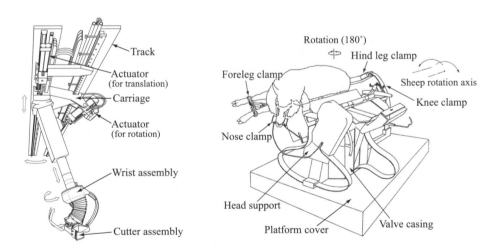

Fig. 1 Wool shearing robot and sheep mounting device

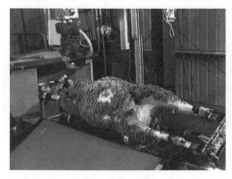

Fig. 2 Sheep shearing robot   Fig. 3 Wool clipper

end-effector to shear the wool with minimal overlapping, no uncut wool and no double shearing. The computer calculates the most efficient shearing procedure for individual sheep to achieve these objectives. It also has a learning function to update the body shape information in its memory as it performs body shape recognition during shearing. It takes 24 minutes to shear over 95 percent of a sheep's wool. A faster process is required (Movie).

To improve the shearing speed, Merino Wool Harvesting Pty Ltd. has developed a system using a different type of platform as shown in Figure 4. A sheep is mounted on this platform as if it is riding a bicycle. The positions of metal ankle braces to immobilize the sheep are adjustable according to the size of the sheep. To keep the sheep calm during the operation, an electrode is fitted to the mouth and the head clamp has 'blinkers' so that the sheep does not see the robot. The two robots are six-axial **articulated type**[1-3.6] arms which shear the left and right sides of the sheep respectively from the head to the buttock starting from the ridge of the back. This method imposes less stress on the sheep as it maintains a natural posture without being turned over and around. It has also achieved a faster shearing speed of 1.5 minutes, which is about 1/5th of the human shearing speed. However, it only shears the outside of the body and the inside and the abdomen area must be sheared manually.

Fig. 4 Sheep shearing with two arms

3.23 Milking

Milking dairy cows is the most labor intensive work within the livestock barn operation and workers must be on duty for many hours as cows must be milked at least twice a day at regular times. For this reason, milking robots have been developed and put in practical use in both Japan and overseas.

The milking method used by the **milking robot** is based on the conventional milking machine (milker). Four teat cups with stainless steel outer cylinders and rubber inner cylinders are attached to the teats. The cups stimulate the teats and pump out milk by repetitive periodic suction and pause using negative pressure created by an external vacuum pump. The periodic switching is actuated by a pulsator valve.

Research on milking robots was taken up by the Japanese Ministry of Agriculture, Forestry and Fisheries early.[58] Commercial models have been developed in Europe and Japan and are already in practical use in Japan. The currently available milking robots are for cows in free-stall housing (not tied). Cows voluntarily move into the milking stall for a feed and milking is carried out there. In the milking stall, cow recognition, teat washing, teat cup attaching, milk quality and quantity checking, teat cup removal, teat dipping and cow ejection from the stall as well as data management by computer are performed.[59, 60] Teat detection is done by either a laser beam, ultrasound or photo-interruption sensor. A robot arm attaches the teat cups to the detected teats.

Figure 1 shows the milking robot that has been introduced widely to Japan. Figure 2 shows the a cow inside this robot. The cow enters through the gate on

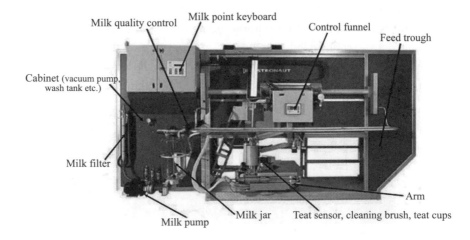

Fig. 1 Milking robot (Lely's Astronaut, Cornes AG, Holland)

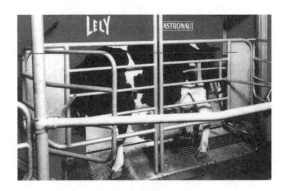

Fig. 2 Cow in milking stall   Fig. 3 Milking by teat cups

the right side of Figure 2 and exits through the gate on the left side once milking
is finished. The **laser scanner**[(I-2.4)] sensor detects the position of the cow's teats
inside the milking stall. The robot arm attaches four teat cups as shown in Figure 3
and starts milking.

Since **stanchion stall barns** are more common in Japan, research on milking
robots for cows in tie-stall housing is undertaken[61(II-1.13)] (Figure 4). The robot

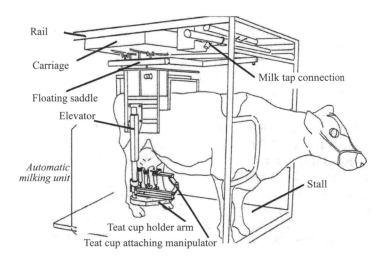

Fig. 4 Milking robot for cows in stanchion stall barn[61]

must be transported to the cow for milking in the stanchion barn. It uses the same procedures for teat detection and teat cup attachment as the free-barn milking robot but it must stop the cow from getting out of position by either immobilizing the cow's body or linking the **milking unit** with a floating saddle which moves with the cow's body.

Plant factories are broadly categorized into two types: one using natural sunlight and the other using controlled artificial lighting. Large-scale facilities using sunlight are operating on a commercial basis all over Japan. Products such as *kaiware* (radish sprout), *mitsuba* (Japanese honewort), leafy lettuce, tomatoes, strawberries and various seedlings are currently being produced in plant factories.

Keywords for the requirements of the next generation agricultural production system include food safety, high quality and stable supply. Discrimination of domestic products from cheap imports and increased self-sufficiency are urgent issues and therefore high expectations for **plant factories**[1-3.2] as one of the solutions. Corporations are beginning to enter the farming business and attempting to secure stable supply of certain crops. Figure 1 shows one example of **plant factory using sunlight**. This facility in the town of Sera in central Hiroshima began tomato production in its 3 ha hydroponics factory in July 2001 and expanded the factory to 8.5 ha in 2005. This was a large-scale corporate style farm operation that had never previously been seen in Japan. It had a state-of-the-art cultivation facility and adopted a new style of farming business with its own production, distribution and marketing systems in partnership with Kagome Co., Ltd. It was incorporated in December 2003.

Each greenhouse is a multi-span type called Venlo or Dutch Light with a width of 8 meters (roof width 4 meters x 2), a depth of 4 meters and a column height of 5 meters. It is fully covered by transparent glass 4 millimeters in thickness. It has hanging gutter type shelves for cultivation. Plants are suspended from beams in the shelves, each of which are formed with no seam for the length of each row (about 64 meters). This system offers a number of benefits. It eliminates the need for workers to squat or bend over, thus reducing their work strain. It is possible to ensure an appropriate inclination for waste water recovery and recycling. Complete separation from the ground means a reduced risk of pest and disease infection. It offers a better growing environment. The introduction of this system began six to seven years ago in the leading protected agriculture country of Holland. The example described here was the first to be introduced to Japan.

This facility is equipped with a complex **environmental control** system. It controls the environmental conditions for plant growth such as temperature, humidity and CO_2 levels by multiple means such as opening and closing roof windows and curtains and temperature control. The irrigation system is a drip tube feed type and uses a nutrient solution mixing device to regulate the **EC** and **pH** levels of the solution. The mixing device blends the recycled solution, water

Fig. 1 Plant factory using sunlight to grow tomatoes (Sera Farm Co., Ltd.)

Fig. 2 Crop management operation using electric motor vehicle (Sera Farm Co., Ltd.)

and fertilizer solutions to predetermined levels of EC and pH for various growth stages. The composition of fertilizer is adjusted by the grower according to the variety and growing condition of the crop.

At various stages of operation, including crop management, harvesting and grading, a range of labor saving equipment is used, including hoist type carts for training and disbudding (Figure 2), harvesting carts, harvesting cart tractors and **pest control robots** (Figure 3). The harvested products are sorted by **grade and class** at the onsite grading facility. Tomatoes are packed in bags or trays, then in cardboard boxes and directly sent to retailers and distribution centers. However, most of these tasks require human assistance.

As above, the world's leading-edge designs and advanced technologies are used for controlling the cultivation environment. However, the current operation requires twice as many workers as those at the Dutch facilities and there are great hopes for automation and robotization.

The leading examples of a **completely controlled plant factory** include TS Farm developed by Kewpie Corporation and the **automated plant factory** developed jointly by Kyushu Electric Power Co., Inc. and Mitsubishi Heavy Industries Ltd. TS (Triangle Spray) Farm uses a spatial configuration spray culture system by which vegetables are grown on triangularly configured panels under high-pressure sodium vapor lamps instead of sunlight and fed by nutrient solution sprayed onto their roots. Plants grow faster under this system because their roots can receive ample supply of water, oxygen and nutrients and, in particular, they can freely absorb oxygen from the atmosphere. The factory can

Fig. 3 Chemical spraying by robot
(Sera Farm Co., Ltd.)

Fig. 4 Cultivation system at TS Farm
(Kewpie Corporation)

produce butter lettuce, leafy lettuce, spinach and herbs of a given quantity and quality at a given time on an industrial scale, unaffected by daylight hours or the weather. The air inside the facility passes through filters which creates a pest and disease free and low bacterial environment to produce completely chemical free vegetables. The adoption of this spatial configuration spray culture system, the artificial lighting system and the automatic temperature and humidity control system has enabled a short production period of about one month from sowing to shipment throughout the year. In summer, night and day are reversed to take advantage of cheaper off-peak power.

The automated plant factory features bar type nursery beds set on a horizontal plane as shown in Figure 5. Interrow spacing is increased according to the growth stage for efficient use of cultivation space as shown in Figure 6. The production process at the factory from seeding to harvesting and boxing is fully automated and human workers are required only to carry out supplementary tasks before and

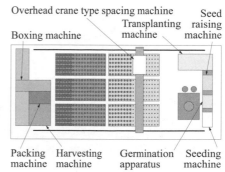

Fig. 5 Cultivation system at
automated plant factory (Kyushu
Electric Power Co., Inc.)

Fig. 6 Facilities at automated plant
factory (Kyushu Electric
Power Co., Inc.)

after cultivation. It has achieved a labor saving of 91 percent compared with its pre-automation operation. Workers seldom enter the production space during the cultivation period and thus the risk of introducing pathogens is extremely low.

As above, computer-based automation of the environmental control inside the plant factory and nutrient solution management are well underway. Also, the development and practical application of automation devices are being considered for various stages of operation, including seeding, seedling raising, transplanting, spacing, harvesting and grading. Some examples of machines for seeding, transplanting and spacing are described next.

While a majority of seeding machines use **coated seeds** for easy handling, some machines can sow naked seeds. However, not all varieties of seed can be used on these machines since there are variations in the shape of individual seeds and of different varieties/cultivars of seeds. A polyurethane cube with a hole in the middle is often used as a seedling raising tray because it is easy to handle when transplanting. Seeds are inoculated either in all cells in a seedling raising tray at once or in a row of cells at a time. With the former method, a template matching the pitch of the cells in a tray is used. The template has holes to capture seeds using vacuum suction and positions itself on top of the cell tray and release the seeds (Figure 7) (Movie 1). The latter type of machine has the same number of nozzles as the number of cells in a row which capture seeds by vacuum suction and drop them in the cells. The tray on the conveyor slides row by row in conjunction with the sowing action. There is a device to shake the seed feeder bucket to facilitate seed capture by the nozzles when handling naked seeds (Figure 8) (Movies 2 & 3).

Seeded trays are watered and conveyed to the germination apparatus. Germinated trays are conveyed to the transplanting machine which uses four pins to grasp each polyurethane cube and transplant seedlings to six holes in each

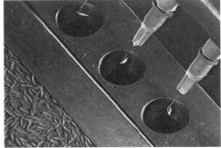

Fig. 7 Vacuum seeding machine using template (Kyushu Electric Power Co., Inc.)

Fig. 8 Seeding machine for naked seeds (Kyushu Electric Power Co., Inc.)

Fig. 9 Transplanting machine   *Fig. 10 Spacing robot (Kyushu*
(Kyushu Electric Power Co., Inc.)   *Electric Power Co., Inc.)*

bar type bed. It uses a **Cartesian coordinate type**[1-3.6] arm. Figure 9 shows the transplanting procedure (Movie 4).

The **spacing robot** (Figure 10) shifts the seedlings to the cultivation beds with wider interrow spacing as they grow larger and consists of a mechanism to lift the cultivation bar and a traveling device to shift it. It hooks its nails to holes in the cultivation bar to lift the seedlings and uses a 3DOF overhead traveling crane to shift them to the next stage. The same machine moves harvesting-stage vegetables to the harvester machine (Movie 5). Butter lettuce and leafy lettuce crops grown under this system suffer no deformity or discoloration by pest or diseases and consumers can eat everything other than the roots without washing. Accordingly, post-harvest processing is not necessary and all that the harvesting machine has to do is to cut off the roots growing below the cultivation bar. After harvesting, the vegetables are weighed and sorted, bagged and boxed mechanically before being shipped (Movie 6).

Agricultural robots handle processed foods as well as living things. This section describes an end-effector developed by Silsoe Technology Ltd., U.K. for handling sticky food and a transferring system between conveyors using air jets.

Conventional open-and-shut type end-effectors have trouble handling jelly, cheese and other very moist food products because their stickiness prevents clean release from fingers or accurate positioning after release. Silsoe Technology Limited has developed an end-effector which is capable of steadily grasping and releasing sticky objects (Figures 1 & 2). This end-effector[62] consists of two fingers which move up and down and a tape attached to the surface of each finger as shown in Figure 3 (Movie 1). One end of the tape is fixed to the end-effector and the other end is wound up by an actuator. The fingers are extended to grasp the target object. When they release the object, they stay closed and the tapes around both fingers are wound up at the same speed. The object stays in the set position and the fingers ascend along the surfaces of the object, eventually releasing it. In other words, the mechanism works as if it is a pair of vertically sliding conveyor belts which sandwich the target object from two sides. Since the fingers ascend after transferring the object to the target position instead of trying to drop it on the right spot, it can achieve accurate positioning. When this end-effector is used horizontally or at an angle, it can handle stacked objects as shown in Figure 4. Besides being able to handle sticky objects, this end-effector has the advantage of being able to perform in very tight spaces as it releases target objects without opening the fingers. It can be used in facilities where strict hygiene management is required if appropriate measures are taken to keep the tapes clean all the time.

The next example is a system for handling (transferring) **processed food**, also

Fig. 1 Cherry fruit handling

Fig. 2 Cake handling

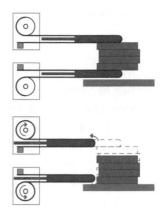

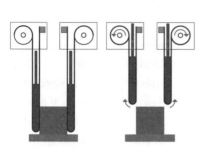

Fig. 3 Mechanism of end-effector   *Fig. 4 Handling of stacked objects*

developed by Silsoe Technology Ltd. (Movie 2). This system uses air jets to change
the traveling direction of processed food at intersections between conveyors[63]
(Figure 5). In the conventional systems of transfer between different conveyors, the
target object is either dropped on to the next conveyor or put on a different track by a
mechanism similar to a railway track switch. However, these methods cannot be used
for high-speed transfer of fragile items such as biscuits. This system uses a device
that emits jets of air from above and below at the intersection of conveyors to change
the direction of flow of processed food. The biscuits traveling on the conveyor from
the left side of Figure 5 are transferred to a crossover conveyor or jump over it and
go straight ahead depending on the strength of the air jet at the junction. Figure 6
shows a floating biscuit. This system is capable of high-speed transfer by changing
the direction of eight biscuits per second at a speed of 1 meter per second. In addition
to transfer between conveyors, it can be used in conjunction with a processing food
inspection system to sort accepted and rejected products at high speeds.

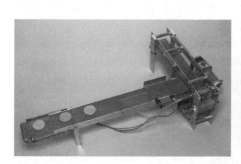

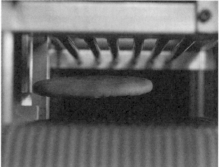

*Fig. 5 Transfer of biscuits between
conveyors using air flow*

*Fig. 6 Floating biscuit at intersection
between conveyors*

Coffee break

Co-evolution of 'human, nature and robot'

Many Japanese people who travel overseas for a long period worry about dietary changes as well as the language barriers they are likely to experience. Although food culture varies from one country to another and it is impossible to say categorically that one food culture is superior to another, Japanese-made food products, especially fresh fruits, are generally high quality, tasty, large and uniform in size and color and some look like works of art when packaged in boxes. When you cut and eat them, a perfect balance of sweetness and acid and a fresh texture will stay with you for a long time. All these are the results of the efforts of producers who have devoted a lot of time and care in the cultivation stage, inspected them with high precision sensors in the grading stage, and taken appropriate quality maintenance measures in the distribution and marketing stage. A majority of products available now are recently improved varieties, not the same varieties as I saw in my childhood. Varieties change with times as if to follow a fashion.

According to authoritative books, fruits tend to have red, orange and yellow colors, which stand out from the green color of foliage, and sweet flesh (flesh remains sour or astringent until maturity) because they are originally organs which have developed to attract animals such as birds and monkeys for preservation of species. Small insects are 'freeloaders' who eat fruits without transporting the seeds afar and plants do not like to attract them except for pollination. For these reasons, petals tend to be very visible to insects and some even reflect ultraviolet light whereas fruits tend to have reddish colors that are highly visible to birds and monkeys but not to insects. It is amazing that some beautiful features of nature have been created by this co-evolution. For example, mimicry of leaf insects is a typical case of evolution for self protection and some slugs even mimic dog droppings so that humans and other animals would not step on them. Such co-evolution is applicable to all living things in the natural world, including humans, but now that humans are about to (try to) co-exist with robots at workplaces, in the community and the natural world, it is perhaps a good time to start thinking about the co-existence and co-evolution of humans, nature and robots. (N. Kondo)

234

4 Vehicle Automation

Practical application of vehicle automation technologies has been advancing rapidly in recent years. Automatic guidance systems for agricultural vehicles are already available in Europe and the U.S.

However, further progress is expected in various fields, including autonomous robots, multi-agent robots and tele-operations. This chapter describes a broad range of topics from the technological components of vehicle automation to the systems of automating specific operations and the latest information on research of next-generation robot systems.

4.1 Drive-by-wire

Drive-by-wire technology for easier operation of steering wheels and other devices is useful not only for manned operation (manual driving) but also as a technology to facilitate automation and robotization of vehicle operation. There are high expectations of further development.

Drive-by-wire[1-4.1] (hereafter called 'DBW') refers to systems for operating aircraft and other vehicles that control rotational and linkage mechanisms using actuators such as electric motors and hydraulic/pneumatic mechanisms which are operated via electric switches (electric wire). These systems replace the conventional operation of rotational and linkage mechanisms in manual operating systems for steering, shifting gears and acceleration/deceleration via steering wheels and levers by the driver.

Driving maneuvers in manned driving which require the use of force can be made easier by the application of DBW technology. The use of sensors to detect the amount of movement enables the driver to execute specified operations with the touch of a key. Since DBW makes it possible to control various devices by electric signals, it makes automation and robotization using electric signal-emitting controllers easier. The current state and future potential of DBW applications are described below using the latest riding type **tractor** as an example.

The main maneuvers for tractor driving include engine start/stop, main and auxiliary gear shift, acceleration/deceleration, **PTO** on/off, PTO gear shifting, implement up/down motion (hydraulic switch), engaging/disengaging the clutch, and braking. Among them, DBW technology has already been applied to engine start/stop and implement up/down maneuvers in widely used relatively small tractors and gear transmission and PTO-related maneuvers in larger high-grade tractors (Table 1).[1]

The DBW control unit converted by BRAIN for the purpose of automation and robotization provides a good case study.[2] A popular relatively small (22.8 kW) tractor was used for this conversion.

The tractor has a fully hydraulic steering mechanism, which is also used on **rice transplanters**. This type of steering is operated by the extension and contraction motion of the **hydraulic cylinder**[1-3.3] (steering cylinder) shown at the top of Figure 1 and the pre-existing manual steering mechanism consisting of manual steering valve and other components is shown on the right side of Figure 1. This mechanism is already a semi-DBW system and can be automatically operated if an automatic steering valve is added in parallel and controlled by electrical signals.

Figure 2 shows the configuration of the automatic control system. It sends signals to magnetic valves via a **PWM** (Pulse Width Modulation)[1-3.3] amplifier to change steering speed and detects the steering angle by a **potentiometer**[1-3.4] for **feedback**

Table 1 Summary of common tractor driving operation mechanisms

Driving operation mechanism	Popular smaller models	High-grade larger models
Steering mechanism	Integral type or orbit-roll type[1] (full hydraulic)	Orbit-roll type[1] (full hydraulic)
Shuttle gear shift	Power-assisted shift	Electro-hydraulic shift[1]
Main gear shift	Power-assisted shift	Electro-hydraulic shift[1]
Auxiliary gear shift	Conventional gear shift	Power-assisted shift
Implement up/down (motion)	One-touch switch[1]	One-touch switch[1]
Implement position	Position lever	Electro-hydraulic position
Throttle	Throttle lever	Throttle lever (some have electronic control[1])
Brake	Direct link	Direct link (hydraulic-assisted[1])
PTO shift	Conventional gear shift	Electro-hydraulic shift[1]
Engine stop	Key-stop solenoid[1]	Key-stop solenoid[1]

Note:

1. DBW or semi-DBW mechanism

control to obtain a given steering angle. The position of the shuttle gear shift, which operates forward/backward/stop travel with a single lever, is controlled by an additional electric motor attached to the lever shaft (the gear shift position is detected by a **limit switch**[1-3.4]). The addition of a motor with an **electromagnetic clutch** to the throttle lever shaft and a hydraulic cylinder to the brake rod end can convert the throttle and the brake into DBW systems or automatically controlled mechanisms.

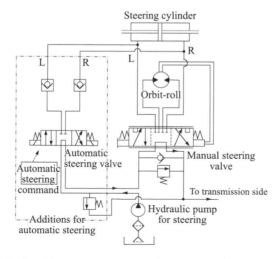

Fig. 1 Full hydraulic steering system of tractor and its automatic control

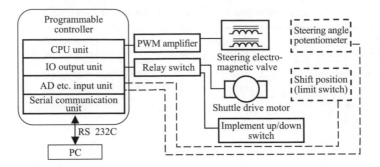

Fig. 2 Block diagram of automatic control tractor

4.2 Operation support systems for farm vehicles (Movie)

> Operation support systems tell the operators in an easy-to-understand manner where they are within a field, how much area they have worked and how to go about the next stage of work. They were researched and developed during the 1990s when the DGPS came into wide use. Commercial models are already available. They help minimize waste and improve efficiency in wide-range or hard-to-track operation.

In farming work carried out by a manned farm vehicle, an operation support system is defined as a system which provides information about the vehicle's current position, its work history and a recommended path to take from that point on a display to help the driver make decisions and operate the vehicle. The **operation navigator for driving support**[3] announced by BRAIN in 2003 and the **Parallel Tracking System** (PTS)[4] commercialized by Deere & Company, U.S., in 2001 are described below as major examples.

(1) **Operation navigator for driving support**: This device is mounted on a farm vehicle such as a tractor. It acquires **GPS**[1-4.1] positional information, directional information and data on the status of the attached implement and displays operational support information such as the current position and work history, and records detailed information about the work being carried out. The operator can easily perform work efficiently and consistently in hard-to-track wide range operations (such as chemical spraying) and night operations by referring to the displayed information.

Figure 1 shows the system configuration of the navigator. The quality of information obtained, displayed and recorded by the navigator depends on the capabilities of the sensors and instruments that gather such information. The navigator's information terminal, an I/O controller (single-board microcomputer with serial ports and A/D conversion ports), has the versatility to accept various sensors according to the quality of required information. The GPS positional information is compatible with a range of formats from the single GPS to the **RTK-GPS**.[1-4.2] It is also compatible with various protocols in terms of the update cycle, speed and data structure of communication data (Movie).

Figure 2 shows an example of the navigator's display for chemical spraying with a spray width of 10 meters (in progress). Idle running tracks appear as dotted lines and worked tracks appear as continuous line segments corresponding to the actual work width (coverage) on the display. The vehicle's current position always appears at the center of the image, which scrolls up automatically in the traveling direction. The image can be enlarged, reduced or scrolled on the touch

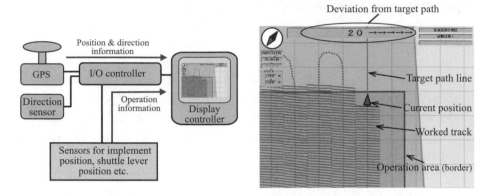

Fig. 1 System configuration of   Fig. 2 Display of Operation Navigator
Operation Navigator (BRAIN)

panel as required. When the vehicle is traveling, the display shows travel guide lines in a grid pattern based on the set work width and indicates a deviation from the nearest guide line (**target path**) parallel to the traveling direction and its direction using numbers and arrows in the upper part of the display. This target path position can be adjusted during operation.

(2) **Parallel Tracking System (PTS)**: This is an operation support system to be mounted on tractors developed by Deere & Company, which displays the target path and the current deviation from the target path on a handy controller (Figure 3). This product is available on the market.

The system obtains positional information from StarFire GPS, which receives data from the GPS information augmentation service run by Deere & Company and NavCom Technology (paid service, available in Japan) and detects the GPS position with a differential of 20 cm or less with no need for a reference station. [1-4.2] Figure 4 shows an example of the controller display. The vehicle driver can perform parallel returning-operations at equal spaces by correcting the course according to the positional data (symbols and numbers) relative to the target path on the display. It can handle a curved line as a target path, which is usually obtained and saved in the system by actually driving on the field.

Commercial products that are even simpler than this PTS (providing only a bar display of a deviation from the target path) are available in the U.S. and Europe for use in **precision farming** (PF). A system which converts deviations from the target path into signals which it uses for automatically steering a tractor is also marketed by Deere & Company as **Auto Trac System**.

*Fig. 3　Controller of Parallel Tracking
System (PTS) (Deere & Company)*

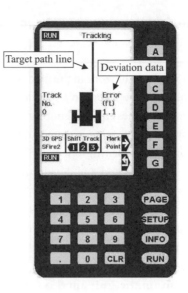

Fig. 4　Display of PTS

4.3 Automatic guidance system – 1 (GDS)

(Movie)

> A majority of work in the field by farm vehicles involves repetitions of straight forward operations in a parallel pattern. Automatic guidance of such vehicles can be achieved by mounting a sensor to detect the heading direction and an automatic steering system to obtain and maintain the target forward direction for each returning-operation. This section describes the systems that have been researched and developed in BRAIN from the late 1980s to date.

The automation of repetitions of straight forward traveling and operation is explained below using the application of the **geomagnetic direction sensor** (GDS) [I-4.2] for detecting the heading direction of vehicles such as **tractors** and **rice transplanters**.

(1) **Automatic guidance system**: This has been developed for the purpose of automating farm work by riding type tractors.[5] Figures 1 and 2 show Trial Vehicle ALVA-I (13.2 kW, 1990) and Trial Vehicle ALVA-II (18.4 kW, 1992) which were converted from commercial model tractors. The steering mechanism, forward/backward/stop switch (shuttle), throttle and implement up/down motion can be automatically controlled by signals from the controller. The sensors used for **automatic guidance** in the final specifications included the tri-axial GDS of FGM-200A by Watson Industries, Inc. and the Sensorex Model 41200 servo type **inclination sensor** (for error correction on inclines).

To detect the heading direction very accurately by a vehicle mounted GDS, it is necessary to cancel the effects of the vehicle's magnetic force and inclination. This system cancels the effect of vehicular magnetism by turning around the GDS-mounted vehicle 360 degrees to obtain a correction value and cancels the effect of vehicular inclination by correcting the GDS output with a directly measured inclination.

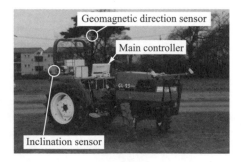

Fig. 1 Trial vehicle ALVA-I (BRAIN)   *Fig. 2 Trial vehicle ALVA-II (BRAIN)*

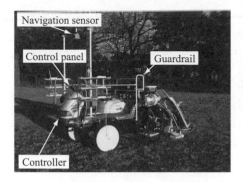

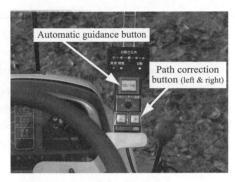

Fig. 3 Auto-steering rice transplanter (BRAIN et al.)

Fig. 4 Control panel of auto-steering rice transplanter (BRAIN et al.)

The system performs its automatic guidance and operation by repeating a series of straight forward traveling and operation for a set period of time (set distance) according to the previously obtained **target forward directions** for the returning-operation paths (obtained in manually driven **teaching** travel) and a 180 degree turn at the path end. ALVA-II is capable of auto-steering travel at a speed of around 0.5 m/s and with a deviation from the target direction of about 0.3 degrees.

(2) **Auto-steering rice transplanter**: This is essentially a **riding type rice transplanter** version of the above automatic guidance system. It was developed for the purposes of reducing the operator's workload and increasing work efficiency and accuracy by implementing automatic steering in the repetitive returning-operation during the manned driving/operation (Movie).[6, 7] The auto-steering rice transplanter has been developed jointly by BRAIN, Japan Aviation Electronics Industry, Ltd., ISEKI & Co., Ltd. and Yanmar Co., Ltd. Its commercial application is currently underway.

The automatic steering system fitted to this rice transplanter (Figures 3 & 4) uses a **navigation sensor**[(I-4.2)] unit to detect vehicle heading direction. The navigation sensor unit is comprised of a GDS, a **gyroscope**[(I-4.2)] and an **acceleration sensor** (to measure inclination) (Figure 5).

The following are the procedures and functions of the automatic guidance system.

① The target forward direction is automatically obtained and set during manual driving prior to the commencement of automatic guidance operation.

② Automatic guidance can be turned on and off as required by pressing the automatic guidance button on the control panel.

③ A lateral deviation from the target path can be corrected during automatic guidance operation by pressing the path correction button on the control panel.

④ A warning sound during automatic guidance operation tells the operator that the vehicle is nearing the end of the path.

The transplanter is capable of operating in automatic guidance mode for a minimum of 30 meters. The lateral deviation from the **target path** after travelling 30 meters is around 5 cm. The lateral deviation may be greater on uneven surfaces or after automatic guidance of a longer distance but the operator can correct the traveling course by simply pressing the path correction button without handling the steering wheel.

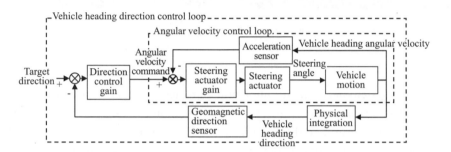

Fig. 5 Vehicle heading control of auto-steering rice transplanter (BRAIN et al.)

4.4 Automatic guidance system – 2 (GPS)

> This section describes two examples of GPS applications: the robot tractor for completely autonomous operation in upland farming developed by NARC in 1999 and the Auto Trac System for automatic steering along the target path commercialized by Deere & Company in 2000.

The **GPS**[I-4.1] is widely used on robots operating outdoors because it does not require any special facilities in the field, it can be used anywhere within the reach of radio waves, and it provides absolute position information.

(1) **Autonomous robot tractor for upland farming**[8] (Figure 1): This vehicle is a 55 kW 4WD tractor which has been converted for autonomous control of the operations listed in the Table below in large-scale upland farming. It achieves navigation accuracy via a relatively low-priced **DGPS**[I-4.2] navigation system with approximately 15 cm accuracy (Leica MX9400N) rather than an expensive dual frequency **RTK-GPS**[I-4.2] and a **Fiber Optical Gyroscope**[I-4.2] (JAE JG-108FD1) for detecting the direction of travel. A **Kalman filter**[I-4.3] is used for filtering and estimation to the positional information from the GPS, the gyroscope[I-4.2] and the speedometer.[I-4.2]

The control software assumes returning-operation with **turns on farm roads** and executes operations such as rotary tilling, levelling, fertilizing and seeding by a simultaneous fertilizer & seeder machine and spraying by a **boom sprayer** (Figure 2) with no human involvement, including traveling between the machine shed and the field. In rotary tilling of a rectangular field 100 meters long and a work width of 2.1 meters, its lateral deviation from the **target path** in straight traveling was 10 cm or less. Longitudinal variation in the operation start position was around 15 cm.

Fig. 1 Autonomous operation of robot tractor (NARC)

Table 1 Main specifications of robot tractor

Navigation system	Position	DGPS (MX9400, accuracy 15 cm)
	Direction	Gyroscope (JG-108FD1)
Tractor (KUBOTA M1-75)	Controlled operations	Steering, three-point linkage, throttle, PTO, engine stop, brake, forward/backward/stop, 4WD on/off, bi-speed turn on/off.
Sprayer (spray width 12 m)	Controlled operations	Boom open/close, up/down, spray pump on/off.
Rotary plow	Width 2.2 m	
Fertilizer & seeder	Controlled operations	Fertilizer distributor open/close.
Controller	Main controller: DynaBook SS3380, Vehicle controller	
Control software	Fertilizing & seeding, spraying, traveling between fields, tilling & levelling.	

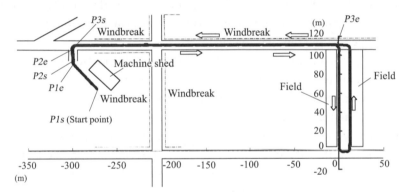

P(i = 1, 2, 3, ...)s: Start point coordinate of linear segment
P(i = 1, 2, 3, ...)e: End point coordinate of linear segment

Fig. 2 Recorded path of robot tractor

(2) **Auto Trac** System:[4] This system, marketed by Deere & Company, detects position by GPS and automatic steering along a preset target path. It is very similar to the Parallel Tracking System described in Section 4.2 and is comprised of a StarFire GPS (Figure 3), a controller that can also be mounted on other vehicles (Figure 4), software and an auto-steering mechanism that can be added to various models of Deere & Company tractors or self-propelled **pest control machines**. As described in Section 4.2, the StarFire GPS receives data from the GPS information augmentation service run by Deere & Company and

Fig. 3 StarFire GPS

Fig. 4 Controller display and
installation kit

NavCom Technology (paid service, available in Japan) and detects an absolute position with a differential of 20 cm or less with no need for a reference station.

Figure 5 shows an example of display on the controller. As in the case of parallel tracking, the target path is set and saved on the system by actually traveling the field or entering the pre-existing position on the path. Reportedly, it can handle curved paths and circling operations as well. Headland turning is performed manually by reference to the target path for the next path.

The users of the Auto Trac System can minimize unnecessary overlaps and reduce the operator's workload to reduce waste of resources and improve efficiency in wide-range or hard-to-track operation.

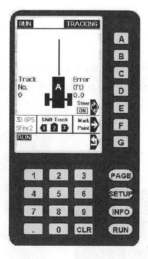

Fig. 5 Controller display

4.5 Automatic guidance system – 3 (vision) (Movie)

> A machine vision-based automatic guidance system was developed by BRAIN in 1995. It uses a machine vision system to detect the boundary between cultivated and uncultivated areas, crops and furrows, and the field boundary as a relative position data, and then utilizes the information for autonomous navigation.

One example of **automatic guidance** based on **machine vision**[I-2.1] is the VSX system developed by BRAIN and Noseiken Co., Ltd.[9(I-4.2)] It was developed by Noseiken for application to lawn mowers. Its **image processor** (VSX) uses an **Hough transformation**[I-4.2] of the original image acquired by monochrome **CCD cameras** in order to detect the boundary line between the mowed and unmowed areas.

The VSX system is an image processor that detects the boundary between tilled and untilled areas (Figure 1) for autonomous tillage operation. It has an **optical filter**[I-2.3] (Figure 2) unit in front of the camera. Selecting an appropriate optical filter according to the weather, soil and vegetation conditions improves its capacity to detect accurately. It divides the original image into 27 x 27 pixels for processing and achieves a sampling frequency of around 30 Hz.

The trial vehicle was a converted 18 kW 4WD **tractor** fitted with a notebook PC-based controller and a **geomagnetic direction sensor** (GDS)[I-4.2] for measuring vehicle heading direction. Its steering mechanism, shuttle gear change, left and right brakes, implement up/down motion and engine fuel cut mechanism are controlled by the controller (Figure 3).

The automatic guidance strategy assumes returning-operation tilling. The first two paths (one returning-operation) are driven manually as a **teaching** run. The

Fig. 1 Original image (left) and Hough transformed image (right) by VSX system

Fig. 2 Optical filter unit of VSX

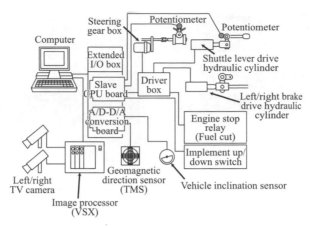

Fig. 3 Component setup of VSX

third and subsequent paths are driven automatically after the required number of paths is keyed into the system. During the teaching run, operational parameters of the VSX such as the camera lens iris and filter selection are set. The run path distance and the forward traveling direction are obtained and saved from the teaching run time and the GDS information respectively. One camera is fitted to each side of the front of the vehicle directly above each boundary line between the tilled and untilled areas based on the work width of the implement. At the start of the third path, the tractor is positioned so that one camera is directly above the tillage boundary line. Once the start button is pressed, the tractor begins automatic operation and stops briefly when it reaches the travel path distance saved in the memory. Then it turns around based on the GDS information and continues automatic operation for the required number of paths along the left and right tillage boundary lines captured by the cameras. It was able to perform automatic guidance successfully when there was a clear brightness difference between the tilled area and the untilled area (Figures 4 & 5).

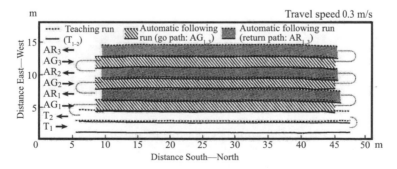

Fig. 4 Result of tillage by VSX system (BRAIN)

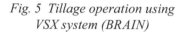

Fig. 5 Tillage operation using   *Fig. 6 Spot-spraying robot*
VSX system (BRAIN)   *(Silsoe Research Institute, UK)*

There are other examples of image processing-based automatic guidance systems. The **spot-spraying robot** developed by Silsoe Research Institute, U.K., is an example (Figure 6).[10, 11] This system uses a monochrome camera for both crop recognition and vehicle navigation. It acquires images of the ground surface obliquely downward in front of the vehicle and discriminates crops from soil by **binarization**[1-2.4] of image data processing and identifies the direction of the crop row by the Hough transformation. Vehicle navigation is executed by the combination of the **inertial navigation** information based on the travel distance information from wheel revolutions and the accelerometer and the vehicle heading direction information from the GDS. The vehicle is a caster type 2WD **transport vehicle** which is automatically guided on the basis of the navigation information and the ridge length, ridge width and ridge number data entered prior to operation. It performs spot-spraying to crops only or herbicide application or mechanical weeding in areas free of crops by distinguishing the presence and absence of crops. The robot performed automatic uninterrupted operation successfully at a speed of 0.7 m/s in an experiment on cauliflower crops. Research on **automatic guidance** tractors based on area-based **stereo vision** has been conducted at the University of Illinois (Figure 7) (Movie).[12]

Left image   Right image   Distance image

Fig. 7 Detection of crop rows by stereo vision

4.6 Automatic following system (Movies)

> This section describes the automatic following system developed by BRAIN in 2000 in which one vehicle (leading vehicle) is driven manually by the operator and the other vehicle (following vehicle) is driven automatically in response to the movement of the leading vehicle.

Automatic following control offers benefits such as labor saving and a substantial improvement of flexibility of farm operation as the following vehicle is self-propelled, not towed.[I-4.7]

The following vehicle of the automatic following system[13] developed by BRAIN is based on an HST 4WD-4WS **remote control** lawn mower (Kubota Corporation). The HST swash plate angle and the front and rear wheel steering mechanisms are controlled with a programmable controller via the D/A board (Figure 1) (Movie 1). The navigation device (LIPS-II) mounted on the following vehicle consists of a flat cable that can be let out in any direction under constant tension. Its end is attached to the leading vehicle. The position of the leading vehicle relative to the following vehicle is calculated from the unreeling direction of the cable β and the unreeled cable length L. The steering angle and other data of the leading vehicle arc transmitted to the following vehicle via this cable (Figure 2).

The automatic following system adopts the L and β values as the target relative position (initial values) between the leading and following vehicles at the moment the automatic following switch on the leading vehicle is turned on. The system controls the HST and the steering angle to maintain the initial values of L and β at all

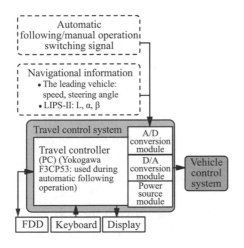

Fig. 1 Component setup of the
following vehicle (BRAIN)

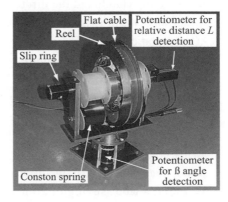

Fig. 2 Navigation device LIPS-II

Fig. 3   Automatic following (BRAIN)

times in response to the movement of the leading vehicle. The automatic following operation is also turned off if the cable is disconnected. In a field test, the following vehicle positioned behind or on the side of the leading vehicle automatically followed while maintaining its relative position accurately and successfully at a speed of up to 1.7 m/s (Figure 3).

The **automatic guidance** inter-row transport vehicle (following vehicle 'RAC') developed by NARC for transporting heavy vegetables is a simplified version of the system developed by BRAIN with two additional functions: automatic inter-row traveling control and remote control. The drawbar, which is the main mechanism of this automatic following system, is connected to the harvester via pins. It is capable of extension/contraction and horizontal swing motions, and the change in the relative position of the leading vehicle is detected based on the amount of extension/contraction and the swing angle. The movement of the drawbar is transmitted to the main traveling clutch/throttle/steering clutch via wires. The main traveling clutch and the throttle are controlled by the amount of extension/contraction of the drawbar and the steering clutch is controlled by the swing angle. RAC automatically travels behind or alongside the leading vehicle while keeping a constant relative distance (Figures 4 & 5) (Movie 2).

The steering clutch is designed to be controlled by an air cylinder as well. It is capable of automatic ridge-following travel and remote control based on control signals from the ridge sensor and the remote control device. In this case, starting and stopping (braking) are executed by simultaneous control of the left and right steering clutches (Figure 6).

Trials for synchronized operation of the RAC and a selective cabbage harvester have found a labor saving of 20 percent compared to the conventional method of operation[14] (Figure 7) (Movie 3). In addition, there has been basic research on relative position detection using a **laser beam** and light reflecting markers[15] as well

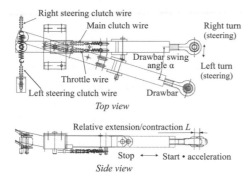

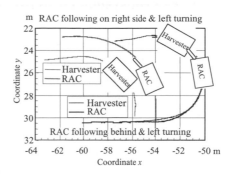

Fig. 4 Drawbar for automatic following (NARC)

Fig. 5 Turning behavior of RAC

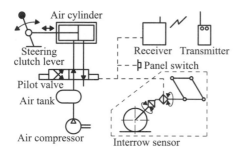

Fig. 6 Components of RAC

Fig. 7 Cabbage harvesting using RAC (NARC)

as studies on the management system for multiple **combine** operations[II-4.7] and the synchronized operation of a **forage harvester** and a **transport** vehicle.[16]

This system was developed at Kyoto University in 1999 to control one or more unmanned combines to follow a manned combine while keeping their relative positions.

The **multi-control system** is designed to allow one operator to control multiple machines for labor saving and better efficiency, not for completely autonomous operation. This system was developed to control multiple **combines.** Each of the computer-controlled unmanned combines (following vehicles) follows the operator-driven combine (lead vehicle) while maintaining a constant distance and lateral transition offset.[17(I-4.7)]

The positional relationship between the lead and following vehicles is shown in Figure 1. Where the coordinates of point P on the leading vehicle are (x, y) on the X-Y coordinate fixed on the following vehicle and the relative heading direction between the vehicles is θ, x is the amount of lateral offset and y is the distance between the vehicles. The values of x, y and θ are obtained geometrically from the distance values d_1–d_4 detected by relatively low-priced **ultrasonic sensors**[(I-3.4)]

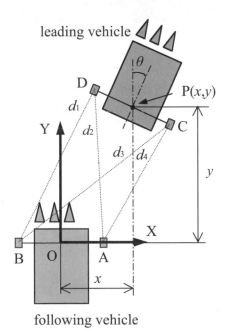

Fig. 1 *Geometry of the leading and following vehicles*

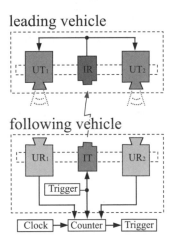

UT: Ultrasonic transmitter
UR: Ultrasonic receiver
IT: Infrared transmitter
IR: Infrared receiver

Fig. 2 *Measurement of relative position between the leading and following vehicles*

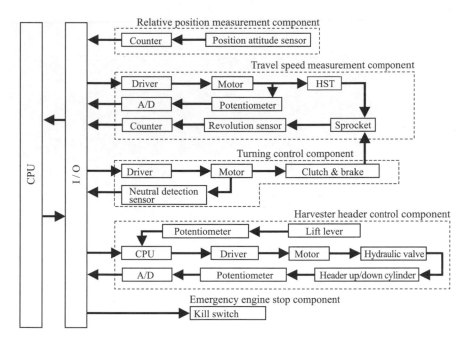

Fig. 3 Measurement and control system of the following combine

fitted to the leading and following vehicles (A–D in Figure 1); distances of A to B and C to D are known.

Figure 2 shows a schematic diagram of relative position detection using the ultrasonic sensors and the **infrared sensor**.[1-3.4] ① A trigger signal (40 Hz) is emitted by the infrared transmitter IT on the following vehicle. ② This signal is received by the receiver IR on the leading vehicle and distributed alternately to the left and right ultrasonic transmitters UT_1 and UT_2. ③ Ultrasonic waves are emitted alternately by UT_1 and UT_2 (the 40 kHz carrier wave modulated on each UT's 20 Hz burst waves). ④ The ultrasonic waves are received by the ultrasonic receivers UR_1 and UR_2 on the following vehicle. The times between the emission from the UTs on the leading vehicle and the reception by the URs on the following vehicle (Td_1–Td_4) are measured. ⑤ The values of d_1–d_4 are calculated from Td_1–Td_4.

Using the above method, it is possible to detect relative positions with an error of several centimeters when the lateral offset (x) is up to 3 meters and the distance between the vehicles (y) is up to 5 meters respectively.

The unmanned following combine is configured so that each component is automatically controlled by the system shown in Figure 3. The travel speed is controlled by **feedback control** using the **potentiometer**[1-3.4] to detect the amount of **HST** lever shift, and the revolving speed of the drive motor is controlled by **PMW**.[1-3.3] The turning (steering) position control and speed control are executed

Fig. 4 Rice harvesting by the leading and following combines

by a drive motor controlled by the computer. Both the combine's travel speed and turning can also be controlled manually.

While the following control based on the relative position is conducted by speed and turning control to maintain the target distance (y) and the target lateral offset (x), the ultrasonic sensors may not be able to detect relative position if sudden heading angle changes occur between the vehicles during headland turning. The leading vehicle therefore stops ultrasonic emission during headland turning to inform the following vehicle of its turning and to make a turn of predetermined actions. An alternative method using a wireless modem to communicate the turning mode is being trialed.

Figure 4 shows rice harvesting tests by one leading vehicle and one following vehicle (Movie). The test using a high precision **RTK-GPS**[1-4.2] to measure the travel paths of the leading and following combines in rice harvesting has found that the following vehicle was able to travel automatically with average deviations of 19 cm/6 cm at the target distance/lateral offset of 3 m/0.6 m and without missing any rice crops.

Farm machinery is a production resource that is expected to play an important role in making farm work lighter, more efficient and more accurate. Research and development of robots that pursue these objectives is being actively pursued all over the world. Hokkaido University and Geo Tec electronics GmbH developed fully autonomous robot tractors in 1998 and 1999 respectively, the current state and issues of which are discussed in this section.[18]

In order to have a **robot tractor**[II-4.8] perform all stages of farm work from tilling and seeding to harvesting, it is necessary to prepare a work plan, including the robot's travel path, in advance as well as converting the hardware of the tractor itself. Figure 1 lists the functions of the robot tractor developed at Hokkaido University. They are grouped into the plan generating function, including travel path preparation, and the autonomous operation function to execute the generated plan faithfully. The work plan generation function uses computer mapping software (Geographic Information System: **GIS**) to make a plan for unmanned tilling, seeding and other operations by the robot. The work plan includes items for the normal switching operation such as forward-backward travel, shifting gears, engine speed, **PTO** and height of **three-point linkage hitch** in addition to the travel path information. These operational items are all pre-set on the GIS. The autonomous operation function in Figure 1 refers to the navigation system capable of operating the robot to autonomously perform farming work based on the generated work plan.

The robot tractor is a converted common farm tractor. Its computer controls the steering, gear shifting, engine speed, implement movement and PTO switching through an RS-232C interface. It uses an **RTK-GPS**[I-4.2] with a 2 cm accuracy, a 20 Hz frequency for positional measurement and a **fiber optic gyroscope** (FOG),[I-4.2] which is commonly used on an aircraft, for directional measurement. Since the **GPS**[I-4.1] antenna is mounted on top of the tractor, inclination of the tractor causes a position error. It also uses an **inertial measurement unit** (IMU) in conjunction with the GPS in order to correct this position error.[I-4.4] Figure 2 shows the robot tractor and the location of the **navigation sensor**.[I-4.2]

As mentioned, the work plan for the robot to operate in the field on its own is prepared in advance. A **navigation map**, in which the positional and path information as well as the operating condition information are saved, is prepared for the map-based guidance system based on the work plan. Coordinate points on the navigation map represent the latitude and longitude of the vehicle position and the 64-bit data showing the vehicle's operating condition. They are defined as shown in Figure 3.[19] The 64-bit data contains information about the tractor's

Functions of robot tractor
— Work plan generation
    ├— Planning using GIS
    │   Tillage, seeding
    ├— Recording and reproducing manually
    │   operated path
    └— Spraying, harvesting, intertillage/
        weeding, farm road traveling
— Autonomous robot operation
    ├— Target path following
    └— Autonomous operation

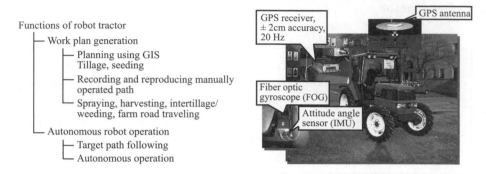

Fig. 1 Functions of robot tractors

Fig. 2 Robot tractor and navigation sensors

operating condition, including the position of the three-point linkage hitch, PTO drive, transmission, travel direction, engine revolutions, total run paths, work width and work status, and are arranged in a bit string as shown in Table 1.

Figure 4 is a flow chart of the robot tractor operation using the navigation map. When the vehicle receives a travel start command, it initializes the FOG to match the GPS position with the FOG direction and reads the navigation map of the path to be traveled. It then selects each navigation point within the map corresponding to the positional coordinates provided by the GPS and determines the controls in each cycle. After performing the required steering control, it reads the data on the

Position

Fig. 3 Data coding of navigation map

Table 1 Coding data

Item	Example	Description	Occupied bit width
Hitch	1	Three-point linkage hitch up/down	1
PTO	1	PTO on/off	1
Shift	3	Gear speed selection	4
Travel direction	1	Forward/stop/backward	2
Engine revolutions	1	Maximum or manual setting	1
Total run legs	5	Set	7
Work width	266	Total run legs on navigation map	10
Work status	1	Interrow width on navigation map indicating whether work has finished or not	1

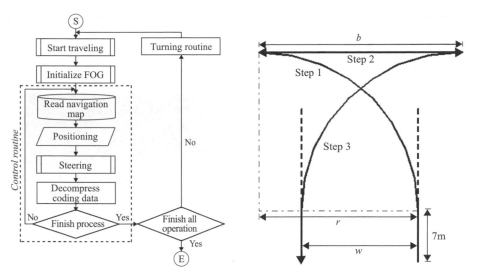

Fig. 4 Flowchart of robot tractor control   Fig. 5 Turning algorithm

navigation point and either finishes the operation if it is at the end of the whole operation or takes a turning action and starts the next travel path if it is at the end of a forward travel path. It uses a three-point-turn, including backward traveling, as shown in Figure 5. This method makes a more complex travel path, but it shortens the distance required for the turn and smoothly guides the vehicle on to Step 3 and the subsequent path by adjusting the reversing distance in Step 2 to offset a variation in the turning radius caused by the ground surface condition in Step 1. The path for the segment to the next path is generated using a spline function in Step 3.[1-4.5] In order to compensate for a change in the ground surface condition, the vehicle enters the next travel path by following the generated path by **feedback control** in Step 3.

All stages of field work, including tillage, seeding (Movie 1), spraying and harvesting, can be automated when the robot is equipped with a map of its travel path. Moreover, it is possible to automate the whole operation sequence completely since the robot can drive out of the machine shed by itself, travel along the farm road to the field, complete the required operation and return to the shed by itself. In other words, there is no need for the farmer to transport such a robot to the field. Figure 6 shows an example of the recorded working path of a robot that successfully traveled on a farm road to the field. The robot traveled four paths while rotary tilling a 2500m² soy bean field. The machine shed was located in the lower right corner of the map. Figure 6 shows the travel path between the machine shed, the field and the four working paths, including turning action. The traveling accuracy of the robot is ± 5 cm, which is far more accurate than a human operator.

Figure 7 shows a spraying operation by the robot tractor (Movie 2). It sprayed chemicals across a 2-hectare sugar beet field using a **boom sprayer** with a spray

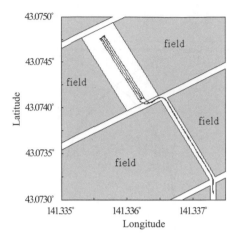

Fig. 6 Recorded working path of
robot tractor

Fig. 7 Spraying in sugar beet field
by robot tractor

width of 18 meters. It was able to complete the spraying without damaging the crops when operating at a speed of 1.5 m/s, which is equivalent to the conventional operation speed. Seedlings in this sugar beet field were transplanted by a conventional manual procedure. While it is self-explanatory that the transplanted crop rows form the path for the next stage of work, this robot tractor can accurately follow the sugar beet rows transplanted in manual operation. This is because this robot tractor is able to record and reproduce the manual drive path as well as being able to operate according to the GIS-based work plan as shown in Figure 1. This enables the operator to record the travel path at the time of seeding or transplanting using the GPS so that the farmer can start using the robot from any work stage.

Robot tractors are being researched and developed all over the world. Geo Tec electronics GmbH of Germany began to market their robot system from the fall of 1999 and reportedly sold almost twenty units by the fall of 2001. The system used on Geo Tec's robot tractor is fundamentally the same as the system developed by Hokkaido University as it uses the RTK-GPS and the IMU as navigation sensors.

In Japan, the practical application of robot tractors is imminent now that it is possible to use the virtual reference station type RTK-GPS and real-time positioning services in the order of several centimeters are available at a low price. However, there are some unresolved issues. One is the issue of how to ensure safety,[1-4.6] both for the operator and in the space around the robot. The awareness of safety is particularly high in the U.S. Even the standard tractors are equipped with a safety mechanism which prevents the vehicle from moving unless the operator is sitting in the seat. Naturally, a higher level of safety assurance is required for a robot which operates without an operator. It is necessary to fit the robot with a multiple-stage and redundant safety

Fig. 8 Obstacle detection using laser scanner mounted on robot tractor

system. Various methods are being proposed, including bumper switches, radar, **ultrasonic sensors**,[1-3.4] laser and **machine vision**[1-2.1] (Figure 8). However, the level of safety that needs to be achieved before a robot can be released on the market is not just a technological question but also a question which requires community consensus. If the robot must take full responsibility in the unlikely event of an accident, it will lead to an enormous cost increase and hinder the progress of robotization. There is a need for discussion and consensus-building with regard to the sharing of responsibility between users and manufacturers and where to draw a line.

4.9 Tillage robot

> A tillage robot is a robot tractor for unmanned rotary tilling. It is capable of working the whole field completely autonomously at around the same levels of operating efficiency and accuracy as manned operation, once teaching data such as the work area and the reference direction are given. This section describes a system developed by a team led by BRAIN between 1993 and 1997.

The **tillage robot** was developed jointly by BRAIN, Kubota Corporation, Japan Aviation Electronics Industry, Ltd. and Hokkaido University with the goal of achieving completely unmanned **tractor** operations in mostly level paddy or upland fields at the same levels of efficiency and accuracy as manned operation (Movie).[20–22]

The robot is mainly comprised of a navigation system, a vehicle system, a controller and operating software. The unmanned robot work is essentially performed according to the conventional work procedures for manned operation. There are three methods of navigation, assuming that part of the navigation system and the operating software can be used selectively according to the situation. The following section describes the vehicle system and controller, which are the common components, and the navigation system and the operating software that are used for the three methods of navigation.

(1) **Vehicle system and controller**

    (a) **Vehicle system**: The vehicle system of the robot is a converted commercial tractor that has to be driven manually from the machine shed to the field. The tractor engine has 23.5 kW of power and an implement for rotary tilling a 170 cm width. Figure 1 shows the converted vehicle system (ROBOTRA) fitted with controllers and sensors for automatic control of various components.

        Figure 2 is a block diagram for control and measurement of the ROBOTRA components. The gear shift position is manually set prior to automatic operation and the engine speed is controlled by two-position switching between a pre-set part throttle and full throttle in order to simplify the control system.

        A wide range of status detection methods are used for each component in order to strengthen self-diagnosis, provide alarms about abnormalities or malfunctions and to serve other functions for improved reliability of unmanned operations.[I-4.6]

    (b) **Controller and safety device**: The controller consists of the main controller and the vehicle controller as shown in Figure 2. A commercial factory computer (NEC) with superior environmental resistance and durability is used for the main controller. A CPU board with an operating frequency of 10 MHz has been manufactured specifically for the vehicle controller.

Fig. 1 ROBOTRA

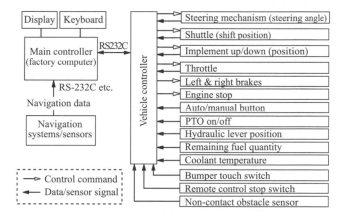

Fig. 2 Measurement and control system of ROBOTRA

ROBOTRA has safety devices ranging from a front bumper switch for emergency stopping if it makes contact with an obstacle to a portable radio transmitter and button switches on the left and right fenders for emergency stopping. In the emergency stop mechanism, a sequence of actions from throttle down, engine fuel cut, brake on and shuttle gear shifting (forward/backward/stop switching mechanism) into neutral is executed instantaneously by the vehicle controller.

(2) **Navigation system and unmanned operation method**

The navigation system adopted for this robot makes concomitant use of information about the position in the field and the heading direction of the vehicle. Three systems are available:

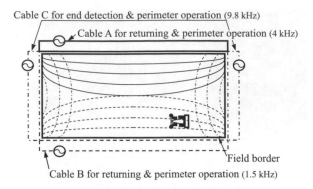

Fig. 3 Cable installation and generated magnetic fields for LNAV system

- Off-the-wire **electromagnetic induction system** (LNAV) with power cable installation around the field;
- **GPS**[1-4.1] system (SNAV) combined with an **Inertial Measurement Unit**;
- Optical measurement system (XNAV) using an automatic tracking type surveying instrument.

The summary and unmanned traveling method of each system is described below.

(a) **Off-the-wire electromagnetic induction system (LNAV)**

   i) **Configuration of LNAV**: The electromagnetic induction system of the LNAV is called the off-the-wire method in which electromagnetic induction is possible at a distance from the installed cable. As shown in Figure 3, electric cables are installed along the borders of a field and sinusoidal current of three different frequencies (1.5 kHz, 4 kHz and 9.8 kHz) is passed through them to generate alternating magnetic fields of different frequencies within the field.

     The LNAV basically determines the vehicle position lateral to the traveling direction by detecting the magnetic intensity within the field and identifying the position in the traveling direction by calculating the travel distance from the rear wheel revolutions. This way, it is possible to generate magnetic field distribution in the whole field for positioning unmanned operations even with a relatively small alternator of several kilowatts. It can perform more accurate positioning near the perimeter of the field by detecting the magnetic field generated by the nearest cable perpendicular to the traveling direction.

     The magnetic field sensor can detect the intensity of the magnetic fields of the three frequencies. One unit is mounted on each side at the front of the vehicle. The distance between them is 10 centimeters narrower than the width of the implement. In addition, a **vibration gyro**[1-4.2] is mounted to detect the vehicle's heading direction information as well as turning angle and to guide the vehicle in sideways movements.

ii) **Unmanned operation**: In unmanned tilling operations using the LNAV, the vehicle works the field except the perimeter by repetitions of straight returning-operations, then works the headlands by traveling along the perimeter (just like the conventional work procedure). The vehicle is first driven manually for one **teaching** run along the long edge of the field and acquires a data string of magnetic field intensity for straight traveling as shown in Figure 4. During the teaching run, the vehicle acquires the magnetic field intensity data string RD0 from the left sensor for the perimeter work to be done later and the data string RD1 from the right sensor for the next adjoining path by reference to the travel distance. In Operation Path 1, the vehicle performs unmanned straight traveling by referring to RD1 with the left sensor while it acquires RD2 for Operation Path 2 with the right sensor automatically at the same time. In this way, unmanned operation using the LNAV system is carried out as the vehicle travels the current path guided by the magnetic field intensity data string obtained during the straight traveling of the previous path while gathering a data string for the next path at the same time.

In headland operation (straight traveling along the four sides of the field), the vehicle travels straight by detecting the intensity of the magnetic field generated by the nearest cable. In the second lap (inside), the vehicle uses the magnetic field intensity data string acquired by the sensor on the opposite side during the first lap. To control 180 degree turning during returning operation, 90 degree turning during headland operation and sideways movement during a transition from a turn to the next straight path, the vehicle uses the output of the vibration gyro to take account of the vehicle heading for smooth vehicle guidance in addition to positioning by magnetic field intensity detection. In unmanned operation using the LNAV, the vehicle is able to reproduce the teaching path almost exactly in a straight run at a speed of 0.5 m/s or so. Its positioning accuracy is considered to be up to ± 5 cm of the target. The data acquisition cycle of magnetic field intensity is up to 0.1 seconds. The AC 100 V, 600 W (200 W per frequency) power source with cables installed as shown in Figure 3 can generate magnetic fields strong enough to guide a vehicle for over 100 meters on the long side and up to 60 meters or so on the short side.

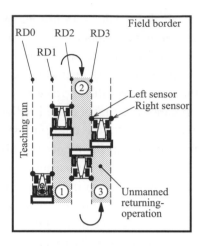

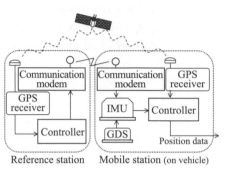

Fig. 4 Unmanned operation   *Fig. 5 Device configuration of*
method of LNAV system   *SNAV system*

(b) **GPS system with inertial measurement unit (SNAV)**

   i) **Configuration of SNAV**: The SNAV is a navigation system based on a
combination of a GPS, an inertial measurement unit (IMU) and a **geomagnetic direction sensor** (GDS)[I-4.2] which measures and outputs
the positional information as coordinates and the heading directional
information about a vehicle. Figure 5 shows the configuration of the
device. It uses a single-frequency interferometric GPS (Motorola),
an IMU manufactured by Japan Aviation Electronics Industry, Ltd.
consisting of a triaxial **fiber optic gyroscope**[I-4.2] (FOG), an accelerometer with a proprietary controller, and a three-dimensional Watson
GDS (in Figure 5).

   ii) **Position measurement**: The SNAV uses a real-time kinematics GPS
by the reference station and the mobile station. It performs static
positioning and bias estimation for the FOG and accelerometer of the
IMU in a stop state at the beginning, and then switches to kinematic
positioning.

      The positional information measured by the GPS has a time lag of
about 3 seconds. The IMU is used in conjunction with the GPS in
order to compensate for this delay because it is able to perform relative position detection and generate more real-time data. The method
of compensation is not explained here due to limitations of space but
this hybrid application gives the SNAV-generated positional data the
real-time quality of IMU measurement as well as the accuracy of GPS
positioning. The heading direction information can be obtained as a

magnetic direction detected by the GDS in addition to a true direction based on GPS data. The IMU can measure an inclination of the vehicle (roll/pitch angle) very accurately. The system corrects for the error with vehicle inclination into the GDS direction data and the GPS position data based on this information from the IMU.

iii) **SNAV performance**: The position measurement of the SNAV has been tested in a stationary state and during traveling at a speed of about 0.5 m/s and found to be accurate to within 5 cm in both cases. The position measurement cycle is up to 0.1 seconds. The heading direction measurement accuracy of up to 0.5° has been achieved after correcting the IMU data based on the GPS output. The direction measurement cycle is up to 0.1 seconds.

The operational range is determined by the data communication distance between the reference station and the mobile station which is approximately 500 meters distant in an unobstructed place.

iv) **Unmanned operation**: Like the XNAV described below, the SNAV is a navigation system which outputs information about the vehicle's position and the heading direction. Its unmanned operation method is the same as that for the XNAV, which is described next.

(c) **Automatic tracking type surveying instrument system (XNAV)**

i) **Configuration of XNAV**: The XNAV observes a target (mobile station) on the vehicle from a range and angle meter (reference station) set in a fixed position outside of the field, acquires the coordinates of the target based on the distance between them as well as horizontal and vertical angles, and transmits the coordinate position data to the vehicle (mobile station) by radio communication. The range and angle measurement instrument at the reference station is a Topcon AP-L1 **total station** which automatically tracks a moving object. It forms part of the positioning system in Figure 6. The heading direction information is measured by a Watson three-dimensional GDS and measurement accuracy is enhanced by correcting for the error generated by vehicle inclination (an **inclination sensor** is mounted on the vehicle) and cancellation of the vehicle's own magnetic force.[1-4.2]

ii) **XNAV performance**: The positioning accuracy of AP-L1 is up to 2 centimeters when the mobile object is around 500 meters away. Since the XNAV takes about 0.2 seconds to transmit data from the fixed station to the mobile station, the accuracy of the positional information received by the vehicle traveling at a speed of 0.5 m/s or so is around 12 centimeters. Because the part of the error stemming from the time delay (10 centimeters) can be corrected based on

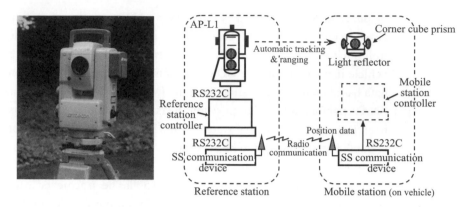

Fig. 6 AP-L1 and measurement system of XNAV system

change in the positional information over time, the positioning accuracy of the XNAV is estimated to be up to 5 centimeters. The measurement cycle for positional information is 0.5 seconds. The heading direction information is estimated to be accurate to within 0.5° when traveling.

The adaptable range is determined by the auto tracking and range measurement distance by AP-L1 or the position data communication distance. It has been confirmed that auto tracking, positioning and data communication can be performed properly in an unobstructed field when the traveling tractor is about 500 meters away.

iii) **Unmanned operation**: Unmanned tillage using the SNAV or XNAV system is basically carried out by repetitions of straight returning-operations of the field except the perimeter followed by headland operation along the perimeter as in the case of the LNAV system. A field coordinate system with axes parallel to the long and short side of the field is set up for the positional information to be acquired by the navigation system so that **path planning** and the vehicle guidance algorithm in the operating software can be simplified.[1-4.5]

Figure 7 is the main flowchart for the operating software developed for unmanned operations by the robot. The software is broadly divided into a planning component and a vehicle control component. The task planning component consists of the teaching module and the task planning module. The vehicle control component mainly consists of the returning-operation module and the round operation module which carry out the actual unmanned operation. The teaching module acquires information regarding the field dimension and work direction. The positional and directional information about the travel

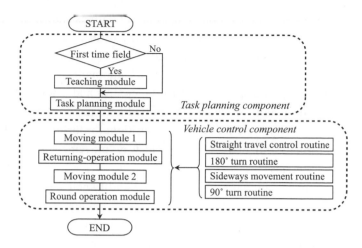

Fig. 7 Main flowchart for unmanned operation

path is acquired by manually operating the robot for one lap along the field perimeter. The teaching run needs to be performed only once for each field when the robot is introduced for the first time.

The task planning module first sets the headland tilling path by determining the work overlap width and the number of perimeter runs based on the field dimension information acquired in the teaching run. It then subtracts the headland operation area from the total field area to acquire the returning-operation area and determines the number of travel paths and the travel path based on the target work overlap width. Figure 8 shows an example of path planning.

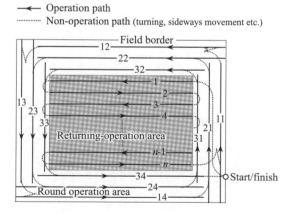

Fig. 8 Path planning for unmanned operation

Unmanned operation along a preset path as shown in Figure 8 is executed by the idle run module, the returning-operation module and the round operation module of the vehicle control component. Each of these modules consists of a combination of various routines for straight travel control, turning control and sideways movement control, details of which are omitted here due to space limitations. The control cycle for acquiring positional information and steering control in straight traveling and sideways movement is basically about 0.5 seconds.

Since it is important to ensure reliability and safety of unmanned operation in the actual work site, a self-diagnosis function to check that each component is working normally prior to operation and an abnormality alarm function to detect errors and determine and execute an appropriate response have been developed and incorporated in the operating software.

(3) Operating performance of tillage robot

Test results of a comparison of the efficiency and accuracy of the unmanned XNAV-based tillage robot and the manned operation (by ROBOTRA, driven by a skilled operator) under the same field conditions is shown in the Table below. Figure 9 shows the robot in operation.

(a) Operation efficiency

Based on the results shown in Table 1, the machine efficiency (the operating efficiency of the machine) was around 25 minutes per 0.1 ha of land in both the robot operation and the manned operation. The efficiency level is likely to depend on the skill of the operator in the case of manned operation and the performance of the operating software, including path planning, in

Fig. 9 Robot operation (rotary tillage) (BRAIN et al.)

Table Result of operation performance test

| Navigation system | Rice paddy field at BRAIN farm (after harvesting) | |
	XNAV	Manned operation
Field lot size (ha)	50	50
Total time (minutes)	142.1	126.3
Machine operation time (minutes)	128.5	122.9
Operation (tilling) area (a)	49.7	49.1
Work speed (meters/second)	0.5	0.5
Machine efficiency (min/10a)	25.7	24.6
Operator efficiency (min/10a)	2.7	25.3
Operator work time ratio (%)	9.5	100
Untilled area (a)	0	0.4
Total tread-on distance (meters)	67.4	13.1
Straightness of travel (cm)[1]	2.7	19.1
Parallelism in straight traveling (°)[2]	0.01	0.44

Notes:

1. The standard deviation of lateral deviation of straight travel trajectories in all returning-operation legs (Straightness).

2. The angle of the regression line of straight travel trajectories to the long perimeter line of the field lot (Parallelism).

the case of robot operations. Since the operator in the manned operations was highly skilled, it is possible to conclude that the performance of the operating software is as high as that of a skilled operator.

The operator efficiency and the operator work time ratio (the ratio of the time for which the operator's attendance is required to the time required for the whole operation) are indicators of the effectiveness of unmanned operations. The effectiveness of unmanned operations is higher when these values are smaller than those of manned operations. The test result shows that these values for robot operations are about one tenth of those for manned operations, which means that unmanned operations have an extremely high labor saving effect.

(b) **Operation accuracy**

The robot surpassed the skilled operator in terms of the untilled area (the left out area) and the straightness of travel. These are indicators of the robot's control performance such as the accuracy of the navigation system and the vehicle positioning accuracy based on navigational information. The test result indicates that the robot's control performance is sufficiently high.

The total tread-on distance (the amount of previously tilled area which was treaded on by the wheels during headland turning and sideways movement), however, was higher in robot operation than manned operation. This is probably because the robot performed sideways movement after turning

Fig. 10 Robot operation (soil   *Fig. 11 Robot operation (wheat*
puddling) (BRAIN et al.)   *seeding) (BRAIN et al.)*

and 3-point-turns less sharply and treaded on more land while attempting to achieve the target turn. In order to reduce the tread-on distance, the robot needs improved vehicle guidance procedures for more efficient turning, sideways movement and 3-point-turning operations as well as a review of the steering control system to enable more speedy steering.

While this section has explained some aspects of the specifications, configuration and functions of the tillage robot and the performance of robot operations, the evolution of the tillage robot continues in the form of extended applications (Figures 10 & 11) and the use in non-rectangular fields. The tillage robot has not reached the practical application stage due to factors such as the set-up costs and safe use in an open field but it is nearing that stage as the costs of navigation systems and controllers have dropped substantially over the last several years.

NARC developed a rice-transplanting robot in 2003. The robot uses a dual frequency RTK-GPS and an attitude control device (fiber optical gyroscope) to measure its position and travels autonomously under computer control with a position measurement accuracy of ± 3 cm and a guidance accuracy of 10 cm.

The **rice-transplanting robot** developed by NARC[23] is based on a commercial **riding type rice transplanter**, which has been modified to perform computer-controlled **automatic guidance** with the addition of DC servo motors for operating the throttle, the gear transmission (**CVT**) and the implement clutch, proportional hydraulic valves for steering control, and hydraulic electromagnetic valves for operating the left and right brakes, clutches and the implement up/down motion. All measurement and control procedures are performed by a single control computer via the input/output interface (RS-232C etc.) (Figure 1).

The **rice transplanter robot** uses an **RTK-GPS**[I-4.2] for positioning which acquires correction data from a reference station near the field via a low-power wireless modem. Since the **GPS**[I-4.1] antenna is fitted to the highest point of the transplanter to minimize obstruction, the ground level position information tends to be affected by the inclination of the vehicle. The heading direction of the vehicle is therefore measured by an uniaxial **fiber optic gyroscope**[I-4.2] and the roll and pitch angles are measured by a triaxial fiber optic gyroscope (FOG) **attitude measurement device** to correct the positional information at the antenna by

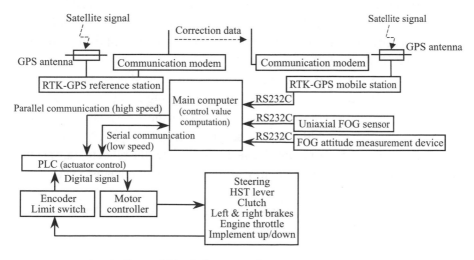

Fig. 1 Control block diagram of rice-transplanting robot

Fig. 2 Transplanting by robot

reference to the ground information.[1-4.4] Since there is a time lag in data output of the GPS receiver, time lag correction is calculated to keep the measurement error within a ± 3 cm range.

The engine speed and operating speed are automatically set based on the position data. In straight traveling, the steering angle and steering speed are determined based on a lateral positional deviation and a heading directional deviation from the **target path** and control signals are sent to the hydraulic valves. At the end of the field, the robot raises the implement, reverses for a short distance, makes a U-turn, enters the next travel path and continues the rice transplanting operation (Figure 2). The control software keeps the transplanting robot traveling within an error range of ± 10 cm from the target path (Figure 3).

The positions of the four corners of the field are measured in advance and saved in the computer as initial data. Operation **path planning** is automatically carried out from the field size and the transplanter width when the robot is set up in the field and the program begins. Although rice seedlings must be supplied manually, the use of **long-mat type hydroponic rice seedlings** enables the robot to transplant up to 0.3 ha of land at a rate of 20 minutes/10a (0.1ha) without replenishing seedlings (Figures 4 & 5) (Movie).

Unmanned rice transplanting operation methods are currently being studied at Mie University[24] and Tottori University.[25]

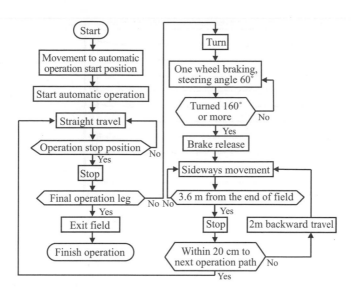

Fig. 3 Control flow diagram

Fig. 4 Transplanting robot operation using long-mat type seedlings

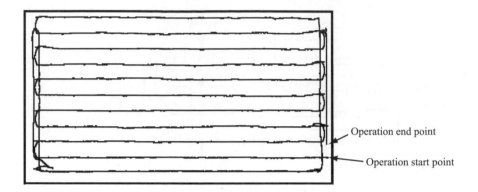

Fig. 5 Trajectory of robot transplanting

4.11 Weeding robot for rice paddy field

In 2000, NARC for Kyushu Okinawa Region developed a 4.9 kg automatic guidance robot which removes weeds by agitating the 1–2 cm depth of soil with circular brushes between plant rows in paddy fields. Its weeding action has an added effect of ridging to plant rows.

Due to increasing interest in reduced-chemical and chemical-free farming in recent years, the development of unmanned mechanical weeding machines is highly anticipated. This **weeding robot**[26, 27] is designed to remove young weeds which are difficult for general weed control machines to remove.

The robot (Figure 1) is 30 cm in overall length, 23 cm in overall width, weighs 4.9 kg and is powered by a **DC motor**[I-3, 3] using rechargeable Ni-Cd batteries for RC (7.2 V). The traveling section has paddle type wheels at the center which propel the robot at a speed of 1 to 2 meters per minute and makes a turn by varying the rotation rates of the left and right wheels. The traveling section uses a **micro switch**[I-3.4] to detect rice plants and a pressure-sensitive switch (bumper switch) to detect ridges. It is controlled by a single board micro computer for **automatic guidance**. The operation section consists of a pair of circular synthetic brushes (110 mm in diameter, approximately 125 rpm) driven directly by motors (Figure 2).

The robot can operate in paddy fields with water 3–5 centimeters deep by agitating the 1–2 centimeters depth of soil with the weeding brushes. This weeding action creates a furrow between rows. In subsequent operations the robot can use the furrow as a guide for more stable traveling. It also has a ridging effect (Figure 3). It performs simultaneous inter-row and intrarow weeding by removing weeds in the inter-row space with the brushes and burying weeds between crops by covering the soil.

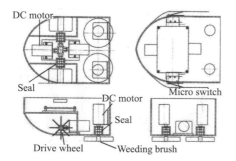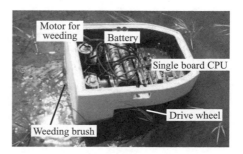

Fig. 1 Weeding robot for rice paddy field (NARC for Kyushu Okinawa Region)

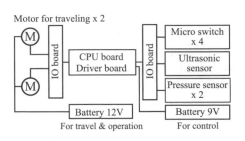
Motor for traveling x 2

Paddy field after weeding: drained for photography

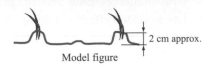
2 cm approx.

Model figure

Fig. 2 Components of weeding robot   *Fig. 3 Ridging by weeding robot*

In a field test of this robot, millet seeds were scattered in the field (500 seeds/m^2) and the amount of weed growth for the following three test lots were compared: Lot A was weeded by the robot, Lot B was weeded by a combination of the robot and liquid mulch, and Lot C was not weeded. Robot weeding was carried out daily from the 7th day to the 30th day after transplanting. A survey carried out four weeks after transplanting found that the number of millet plants in Lot A was about one half of that in Lot C and the number in Lot B was about one tenth of that in Lot C (Figure 4). The effect of mechanical weeding has been more marked in the case of the weed '*Konagi*' (Figures 5 & 6). The same survey found that the average height of millet was 26 cm in Lot A, 9 cm in Lot B and 50 cm in Lot C. The yields of Lot A and Lot B were approximately 20 percent higher than that of Lot C. The robot was capable of about two hours of continuous operation using the mounted batteries.

The test revealed that the robot had some trouble in turning before going into the next path or performing straight traveling smoothly in the first weeding operation

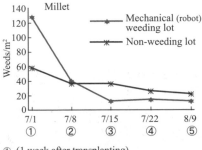

① (1 week after transplanting)
② (2 weeks after)   ③ (3 weeks after)
④ (4 weeks after)   ⑤ (Weeding finish)

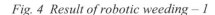
Note: The weed number in week 1 is prior to weeding start.

Fig. 4 Result of robotic weeding – 1   *Fig. 5 Result of robotic weeding – 2*

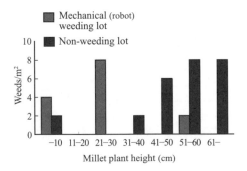

in some cases because the performance of the micro switch for rice plant detection was unstable when detecting soft seedlings soon after transplanting. This issue needs to be resolved in the future.

Fig. 6 Result of robotic weeding – 3 (Left : non-weeding Lot C, Right : robot-weeding Lot A, 30 days after transplanting)

4.12 Mobile mechanisms for greenhouse robots (Movie)

> The mobile mechanism has a decisive impact on the performance of greenhouse robots since it has to be compact with a small turning radius yet structurally strong enough to support the weight of arms and end-effectors. This section describes the systems developed by Ehime University (1991), National Institute of Vegetable and Tea Science (2003) and BRAIN (2002).

Various R & D projects are underway for greenhouse robots, including harvesting robots for tomatoes, eggplants, cucumbers and strawberries. The development of unmanned pest control robots is also progressing toward practical use. The structure of the upper body of harvesting robots which include operation units consisting of arms, end-effectors and vision units, were described in Chapters 2 and 3. The mobile mechanisms, the lower body, of these greenhouse robots may have different features depending on the type of crop or work they handle but a majority of them are structurally similar. The traveling mechanisms of some of the transportation vehicles under development can be used for robots as well.

This section describes a few of the mobile mechanisms already developed or under development, namely, a **battery-car in greenhouse** (originally 'unmanned transport vehicle for use in greenhouse') developed by Ehime University et al.,[28, 29] the mobile mechanism of an **eggplant harvesting robot**[II-3.13] developed by the National Institute of Vegetable and Tea Science et al.,[30] and a **utility battery carrier in greenhouse** (originally 'small operation vehicle for vegetable production') developed by BRAIN,[31] and discusses various requirements and functions that are common to these mechanisms. The unmanned **pest control robot** for greenhouse is dealt with separately.[II-4.13]

(1) **Development examples**

    (a) **Battery-car in greenhouse by Ehime University et al.**: a four-wheeled, independent rear-wheel-drive, unmanned (automatic) **transport vehicle** driven by two 12V-150W geared motors (Figure 1). It is fitted with a single battery of 12 V (36 Ah).

        Its **automatic guidance** system incorporates three modes of control: **attitude control** to avoid driving onto a ridge, **pivot turn mechanism** control to make a turn at the end of a ridge, and **mobile control** to automatically move to a preset position.

        Attitude control is performed by purely mechanical means. Because the front axle is fitted to the frame of the vehicle via the rotary steering shaft supported by two ball joints, the front axle rotates when one of the front wheels is about to drive up a ridge (a skewed turn against the rear axle)

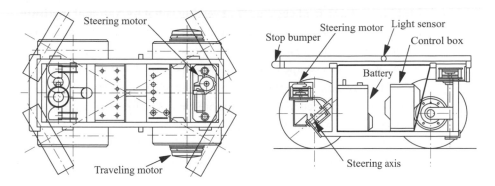

Fig. 1 Construction of battery-car in greenhouse

so that the front wheel descends the ridge slope. For turning at the end of ridge, the independent front and rear wheel motors turn the front wheels in the shape of the character '〉' and the rear wheels in the inverted form of that character by at least 60° to turn-on-the-spot with a turning radius of 440 mm.

Control for automatic transport to a given position is performed based on recognition of its own position by detecting strategically placed reflectors using reflective light sensors mounted at three places on the vehicle. The positions of the reflectors are saved in the micro computer ROM as map information.

Its 150 W travel motors can perform normal and reverse rotation and speed control by **PWM** (Pulse Width Modulation)[1-3.3] and travel at a maximum speed of 50 cm/s or so. It is capable of traveling at 40 cm/s carrying a maximum load of 80 kg. Its hill-climbing ability is around 12° and its continuous operation with a fully-charged battery and a 52 kg load is about 5 hours or 4.8 km movement. Figures 2 and 3 show the vehicle carrying out transport operation and pest control operation respectively (Movie).

*Fig. 2 Battery-car in greenhouse
(used as carrier)*

*Fig. 3 Battery-car in greenhouse
(used for spraying)*

(b) **Mobile mechanism of eggplant harvesting robot by National Institute of Vegetable and Tea Science et al.**: This is the traveling section of a robot system for harvesting eggplant fruits grown on V-shaped trellises in greenhouses. V-shaped trellis training is considered to be a suitable cultivation system for robotic harvesting since eggplant fruits set on the main branches trained into a V-shape are harvested first and fruits on side branches are harvested as they set.

Figure 4 shows this robot system. Its traveling section (mobile platform unit) at the bottom measures (approximately) 1450 mm long, 640 mm wide and 610 mm high. The mobile platform is based on a commercial model electric work platform (ISEKI & Co., Ltd.: BW-400S) designed to travel on a pair of pipes (48.6 mm in diameter) that are set 425 mm apart between ridges. It obtains electric power from a cable on a pulley which moves along a wire installed in the overhead space inside the greenhouse.

The mobile platform controls its traveling direction using **limit switches**[1-3,4] fitted to the front and rear of the vehicle to detect the markers inside the greenhouse and a **rotary encoder**[1-3,4] to detect wheel revolutions. The method of transport on headland passageways is currently being studied.

(c) **Utility battery carrier in greenhouse by BRAIN**: This vehicle assists manual operation in high places and heavy material handling inside the greenhouse. Figures 5 and 6 show prototype models. The targeted specifications include a carrying capacity of 100 kg, a loading platform elevation capacity and a traveling section that is highly maneuverable in a confined space. Two 12 V – 17 Ah batteries are mounted as its power source and independently drive four wheels called mecanum wheels.

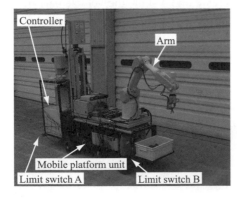

Fig. 4 Eggplant harvesting robot

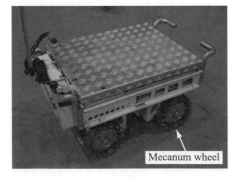

Fig. 5 Utility battery carrier in greenhouse

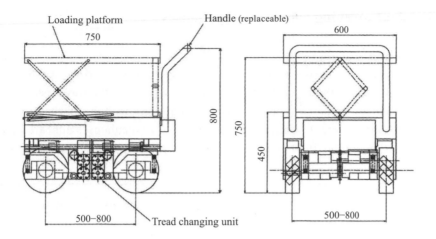

Fig. 6 Construction of utility battery carrier in greenhouse

The mecanum wheel has a series of evenly spaced rollers around it at an angle of 45° to the wheel axle and the vehicle can move in any direction by controlling the turning direction or speed of the four wheels. Further improvements to its omnidirectional mobility on various road surfaces and the potential for **automatic following** after the operator and remote control are being studied.

(2) **Functions and requirements for mobile mechanism**

The mobile mechanism of a robot is required to have the following structural and functional characteristics.

① It is of a size and structure that allows the robot to travel, turn and change direction in narrow passageways without damaging crops.

② Its structure can support the weight of the operating and vision sections and cope with various reactive forces during operation.

③ It receives electric power supply from batteries or cables.

④ Its functions include straight traveling along a passageway and driving over installed pipes, etc. .

Where it is possible to lay a track, a commercial electric vehicle with cable power supply can be converted into a robot's mobile mechanism as in the case of (b) above but it is necessary to consider a mechanism for traveling on the headland passageways. For a mobile mechanism for free path traveling, the procedure and performance for straight traveling along ridges, turning and change of direction are the key points as in the case of (a) and (c). It is important to consider all factors carefully since too much emphasis on maneuverability and size can result in an inability to support the weight of the upper body or cope with reactive forces during operation.

4.13 Chemical spraying robots in greenhouses

Chemical spraying robots used in greenhouses are broadly divided into the electromagnetic induction type with a guide wire along the travel path and the automatic guidance type that travels along ridges with the mechanism of wheels and casters. Chemical spraying robots in greenhouses appeared on the market around 1990 and were the first vehicle type farm machinery to achieve unmanned traveling and operation.

There was a strong need for unmanned **pest control machinery** due to the high risk of chemical exposure to machine operators during pest control operations (chemical spraying for pest/disease prevention) inside greenhouses. **Pest control robots** for unmanned traveling and operation were developed and commercialized around 1990. **Chemical spraying robots** became the first commercial vehicle type farm machinery capable of unmanned traveling and operation.

Chemical spraying robots generally travel on passageways on both sides and in the middle of a greenhouse (headland passageways perpendicular to ridges/crop rows) and furrow passages between ridges/crop rows. Automatic traveling by chemical spraying robots includes reciprocating furrow traveling while spraying (go-back traveling between ridges in Figure 1) and transfer traveling (to another furrow) on a headland passageway at one end of the greenhouse (vertical path on the left side in Figure 1).

There are two broad types of **automatic guidance** system for chemical spraying robots. One is called the **electromagnetic induction** system under which guide wires are laid along the headland passageway on one end of the greenhouse or in the middle of each furrow and the robot is automatically guided along the laid wires by detecting generated magnetic fields with its induction sensor.[32] [(I-4.2)] The other system, called caster system,[(II-4.12)] uses a caster type wheel mechanism which travels along ridges without driving on to them beyond a certain height. Under either system, a marker such as a post, a magnet or a steel plate (the sensor plate for the ridge end in Figure 1) is placed at the turning point of reciprocating furrow travel and the spraying robot uses a **limit switch**[(I-3.4)] or a magnetic sensor fitted at the front to detect the marker in performing go-back traveling.

Automatic guidance in the electromagnetic induction system is guided by an induction wire along the headland passageway and the turning position to each furrow is found by detecting the sensor plate as shown in Figure 1.

The caster system used in the development of the chemical spraying robot used a platform which traversed on a set of rails laid along the passageways[33] as shown in Figure 2 but this setup cost time and money. Under the current caster system,

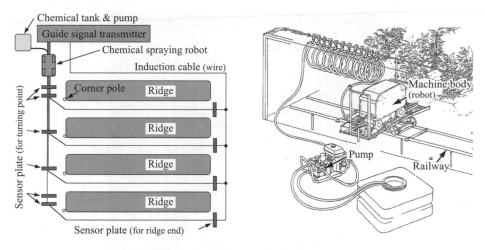

Fig. 1 Installation of electric cable etc. for navigation

Fig. 2 Carrier and railway for lateral movement (around 1990) (KIORITZ Corporation)

either the robot automatically turns and transfers by detecting a magnet or other types of marker placed at the turning point in the passageway or it is manually operated when changing direction and transferring in the passageways. For automatic turning, a pivot shaft extends from the bottom of the robot body, lifts the body and turns it 90 degrees.

Chemical spraying robots have features such as a choice between the two-way spraying mode for chemical application on both paths of a go-back furrow travel and the one-way spraying mode for chemical application on the return path only (high-speed travel on the outward path), and an automatic reeling/unreeling mechanism for the chemical hose to minimize slack (the chemical tank and the pump are stationed outside of the greenhouse).

Commercial models of chemical spraying robots include Shuttle Spray-Car[34] by MARUYAMA Mfg. Co., Inc. and Robot Spray-Car[35] by KIORITZ Corporation, both of which offer both electromagnetic induction and caster types (Figures 3, 4 & 5).

Fig. 3 Unmanned spraying by "Shuttle Spray-Car" (MARUYAMA Mfg. Co., Inc.)

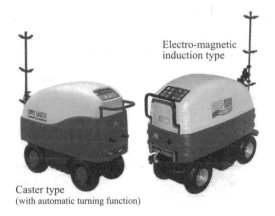

Electro-magnetic
induction type

Caster type
(with automatic turning function)

Electro-magnetic
induction type

Fig. 4 "Shuttle Spray-Car"
(MARUYAMA Mfg. Co., Inc.)

Fig. 5 "Robot Spray-Car
(KIORITZ Corporation)

Pest control robots for orchards were developed by a team led by BRAIN in 1993. They include a cable navigation type which follows guide cables laid along the operating path and a pipe navigation type which follows pipes laid in the orchard.

The cable navigation type of unmanned **pest control machine** (unmanned SS (**speed sprayer**)) and the pipe navigation type of unmanned **pest control machine,** developed by a team led by BRAIN, were put into practical use in 1993 and 1994 respectively. They are useful for preventing chemical exposure to operators, reducing work load and thus labor saving.

The unmanned SS (Figure 1)[36] adopts a navigation system which guides its travel along cables installed underground (less than 30 cm deep), on the ground or in the air (150–200 cm above ground) along the operating paths within the orchard (Figure 2) (Movie 1). A single-phase AC from a transmitter passes through the cables and generates magnetic fields. The unmanned SS is fitted with various sensors, actuators and controllers. Two induction sensors detect the magnetic fields and calculate a difference between the cable and the machine body. This value is applied to the **automatic guidance** control algorithm using **fuzzy logic** to operate the steering control actuator (hydraulic), which navigates the machine.

It is fitted with an automatic stopping device for safety which is activated when the **ultrasonic sensors**[1-3,4] and touch sensors detect an obstacle, when the machine

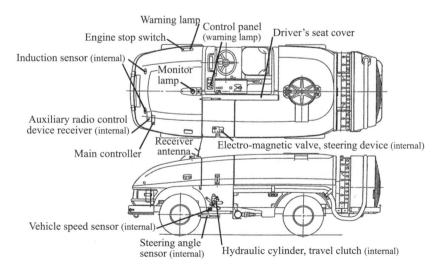

Fig. 1 Schematic diagram of unmanned speed sprayer based on cable navigation

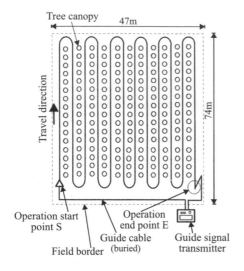

Fig. 2 Example of cable installation   Fig. 3 Unmanned spraying
in orchard

deviates too far from the cable or the cable breaks, or when the **auxiliary radio control device** becomes uncontrollable. It is also equipped with an emergency manual stop device. The cable at the operation finish point is bent at a sharp angle which activates the automatic stop device and halts all functions. There is a **pressure sensor**[1-3.4] for detecting the remaining chemical quantity at the spray pump spout and electromagnetic valves for the rightward, leftward and upward nozzles. In chemical spraying operation during unmanned traveling, the machine performs overall application in rightward, leftward and upward directions between tree rows but when it comes to a turn it detects the steering angle of the front wheels and closes the valve for the spray nozzle on the outside to avoid unnecessary application. When the tank runs out of a chemical solution, air blasting, spraying and traveling stop automatically but chemical refilling must be done manually (Figure 3). The auxiliary radio control device is capable of remotely controlling such functions as the start/stop of the vehicle, the start/stop of the blower fan, and the start/stop of spraying up to a distance of about 150 meters using a **specified low power radio**. Its operating efficiency is on par with that of a manual operation of the same model of speed sprayer and it can be used for conventional manual spraying as well.

The pipe navigation type pest control machine[37] (Figure 4) is a special three-wheeled vehicle with a **rotating nozzle type sprayer** designed for use in orchards with relatively low-height trees, an incline of up to 6° and space for a turning radius of at least 1 meter (Figure 5) (Movie 2). It performs unmanned operation by

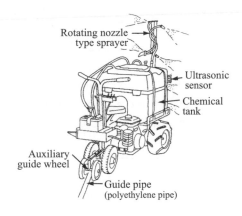

Rotating nozzle
type sprayer

Ultrasonic
sensor

Chemical
tank

Auxiliary
guide wheel

Guide pipe
(polyethylene pipe)

Fig. 4 Schematic diagram of
unmanned pipe following sprayer

Fig. 5 Spraying operation by
unmanned pipe following sprayer

following a polyethylene pipe laid along the strip road within the orchard using a set of auxiliary guide wheels, which determines the steering direction of the dual front wheels. It stops automatically when the auxiliary guide wheels come off the pipe while traveling or when the vehicle makes contact with an obstacle.

Speed sprayers using GPS[I-4.1]-based **navigation sensor**[I-4.2] are currently being researched in Korea.[38]

4.15 Vision-based autonomous hay harvester (Movie)

> The most famous vision-based autonomous hay harvester was developed by a research & development project at Carnegie Mellon University. Its machine vision guidance and turning function use a DGPS and a Geomagnetic Direction Sensor. It successfully performed unmanned mowing of a 40 ha pasture in 1997.

The **autonomous hay harvester** (Figure 1) developed jointly by the Carnegie Mellon Field Robotics Center and New Holland in the U.S. uses a **machine vision**[1-2.1]-based **navigation sensor**[1-4.2] (Movie). It successfully performed unmanned mowing operation on a 40 ha pasture in 1997.

The machine vision used as a navigation sensor is a common RGB color camera which detects boundaries based on **image segmentation** (Figure 2).[1-4.2] It detects a boundary between the uncut area and the cut area and steers the machine so that the detected line is always at the center of the image. Accordingly, this robot system allows the human operator to drive it manually to cut grass on the perimeter of the field first, and then switch to unmanned operation.[39]

The navigation sensor is fitted with a **DGPS**[1-4.2] and a **Geomagnetic Direction Sensor (GDS)** [1-4.2] in addition to the machine vision. Turning positions of the machine are determined based on the DGPS positioning data acquired during field perimeter traveling. The harvester is fitted with one machine vision unit on each side so that it can acquire the image of a boundary on either side of the machine. After detecting the operation end point with the DGPS, the harvester hoists its mower header and enters a turning motion. The harvester turns 180 degrees from the previous travel path using the GDS. It continually searches for the boundary of the next path using the machine vision during the turn. Once it detects the line, it executes vision-based navigation of the harvester in the same way as during

Fig. 1 Autonomous hay harvester

Fig. 2 Machine vision-based detection of line between cut and uncut areas

Fig. 3 Obstacle detection by machine vision

mowing operation and once it confirms using the DGPS that it has reached the operation start point, it lowers the header and starts mowing. The machine repeats this sequence of actions to mow the entire pasture.

While the boundary is detected by a relatively simple process, measures to deal with various problems that can arise in the outdoor environment have been incorporated into the system.[40] One of these is a safety issue involving obstacle detection and collision prevention when a mobile obstacle enters the harvester's travel path. This vision-based autonomous hay harvester uses an algorithm to recognize an obstacle by applying image processing techniques such as spatial filtering to any area other than the cut and uncut areas of pasture. Figure 3 shows an example of obstacle recognition when another harvester has entered the path of the unmanned harvester. Grey areas in the image indicate the areas recognized as something other than the cut and uncut areas of pasture; in this example they are the sky and the other harvester. If the system can assume that the pasture is a plane, it can estimate the distance to the obstacle and take appropriate safety measures such as stopping or sounding an alarm.

There is also a need to deal with the harvester's own shadow. When the sun is behind the harvester, the machine vision captures the harvester's own shadow as shown in Figure 4 (a). The boundary line that falls inside this shaded area cannot be recognized by the color-based image segregation method. A simple amplification of the RGB signal intensity within the shaded area hinders the reproduction of the image of the sunny area. In other words, the color temperature changes in the shaded area, making boundary detection impossible unless an appropriate measure is taken to compensate for it. As a solution to this problem, the system has incorporated a new method of converting a shaded area into a sunny area through modeling the spectrums of direct light and diffuse light. It corrects the appearance of the

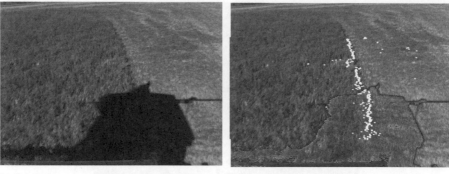

(a) Original image with harvester's shadow   (b) Corrected image and boundary detection

Fig. 4 Grey level correction of shaded area

shaded area in Figure 4 (a) as shown in Figure 4 (b), and then applies the boundary detection algorithm for the sunny area to the corrected shaded area to accurately detect the boundary.

Lawn mowing or grass mowing is a relatively simple operation but it has to be done so frequently that the robotization of this work offers significant labor saving and work load reduction. Some robots are already available on the market. This section describes the robot developed by Fuji Heavy Industries Ltd. in 2001.

Grass mowing (lawn mowing) has to be done quite frequently from spring to autumn and the work in hot and humid conditions imposes great physical stress on workers. Moreover, lawn mowing at golf courses has to be done in the early morning before players arrive and grass mowing at airports has to be done in the middle of the night when there are few flights taking off or landing. For these reasons, there have been attempts to robotize mowing operations.

One of the grass mowing robots already available on the market has been developed by Fuji Heavy Industries Ltd.[41] (Movie 1). This robot is based on a 4WD grass mower (Table 1) with actuators and feedback sensors added to the operation control section, including a steering system, an **HST** mechanism and brakes, and the implement operating device. While the robot is controlled interactively from the control box on the side of the body, the basic operation system layout and procedure remain the same as the original mower so that it can be operated manually as well.

The robot acquires information about its own position from an **RTK-GPS**[(I-4.2)] (a reference station is installed) and uses a **dead reckoning navigation** system consisting of a **rotary encoder,**[(I-3.4)] a **geomagnetic direction sensor**[(I-4.2)] and an **inclination**

Table 1 Specifications of mowing robot

Body size	Overall length	350 cm
	Overall width	184.5 cm
	Overall height	198 cm
Weight		1,220 kg
Drive system		4WD
Engine	System	Water-cooled vertical 4-cylinder diesel
	Displacement	1.335 L
	Output	20.5 kW (28 PS)/2,800 rpm
Speed	Low speed	0–9 km/h
	High speed	0–17 km/h
Implement	Cutting height	3.5–11 cm
	Operation width	153 cm (rotary type)
	Efficiency	53 a/h (at 5 km/h)

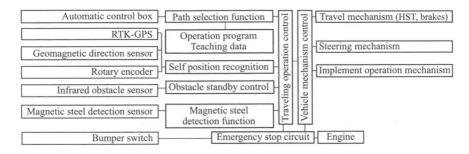

Fig. 1 System configuration of mowing robot

sensor, to acquire supplementary position data until the latest **GPS**[1-4.1] data become available (Figure 1). Where GPS positioning is impossible due to trees and other objects blocking signal transmission, it performs **follow-up traveling** by detecting pre-installed underground magnetic pegs with a magnetic sensor. In other words, the robot uses a hybrid of three navigation systems – **hyperbolic navigation**, dead reckoning navigation and **semi-fixed path navigation** – to achieve a positioning error within a 20 cm radius.

The robot uses a **teaching/play-back** method for automatic guidance operation by which the operator performs manual operation procedures, which are memorized and automatically reproduced by the robot. During **teaching**, the normal manual operation is performed and events data such as recognized positions and operation procedures are recorded. The robot reproduces the recorded data to perform unmanned play-back operation (Figure 2). During unmanned operation, the robot constantly monitors conditions such as loading on the drive system and change in the cutter operating condition according to the grass density and height and controls its speed to flexibly respond to various conditions.

Its emergency stop mechanism is an independent system apart from other control systems which executes forced fuel cut and power-off. The emergency

Fig. 2 Mowing robot operation by play-back

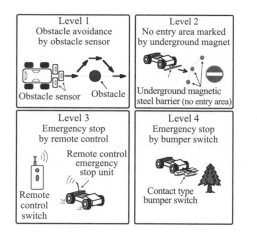

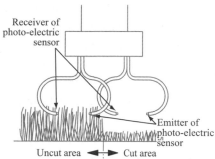

Fig. 3 Ensuring safety of mowing robot   *Fig. 4 Boundary detection sensor*
                                                  on mowing robot

stop mechanism is activated by the emergency stop switch, the bumper switch and the remote controller switch as well as detection of underground magnetic steel markers pre-installed at hazardous areas as a magnetic steel barrier system. It also incorporates a fail-safe design by which an emergency stop action is taken in the case of sensor breakdown, abnormality or harness breakage (Figure 3).

In contrast, the lawn mowing robot developed by Kubota Corporation[42] is based on an HST drive 4WD-4WS **remote control** lawn mower (Kubota AMX3-II). It has a programmable controller which controls the HST swash plate angle and the front and rear wheel steering mechanism through the D/A board. Its navigation system has a boundary line sensor which uses a **photoelectric sensor**[1-3.4] to detect the boundary between the cut and uncut areas of lawn as shown in Figure 4, in conjunction with dead reckoning navigation using directional information from the geomagnetic direction sensor and traveling distance information from wheel revolutions.

To start operation, the robot is driven either manually or by remote control along the perimeter of the area of operation in a rectangular shape and the direction and distance data for this teaching operation are saved. For automatic guidance operation, the robot circles the lawn area from the outer perimeter toward the center based on boundary information from the boundary detection sensor and its memorized operation and distance information. It can achieve a small turning radius by the anti-phase 4WS function when turning and travel along the boundary line without changing traveling direction by the use of the same-phase 4WS (crab steering) during mowing operation. In addition to the above, lawn mowing robots using GPS as the main navigation system are manufactured by Deere & Company of the U.S.A.[43] (Movie 2).

4.17 Fertilizing robot for steep slopes

A robot for safely and efficiently broadcasting fertilizer on steep grassland was developed by the National Institute of Livestock and Grassland Science in 2001. It is a relatively low-cost system using a geomagnetic direction sensor as the main navigation sensor.

In applying fertilizer using a **broadcaster** (around 15 meter application width), the boundary between the fertilized area and the unfertilized area is not clearly visible in most cases. The conventional application operation tends to suffer from accuracy problems such as missing and overlapping because it must rely on the operator's 'guess'. Moreover, there are problems associated with irregular shapes of sloping grassland as shown in Figure 1, for example, such as difficulty in setting an optimum travel path and serious accidents caused by vehicle rollovers. The **fertilizing robot for steep slope fields** in Figure 2 (nicknamed "Kametarō") was developed at the Mountainous Grassland Branch of the Grassland Research Station for the purpose of improving operational accuracy, saving labor and ensuring safety.[44]

This robot is a **crawler** vehicle with radio control capability designed with steep slope operation in mind. Its left and right **HST** swash plate angles and brakes are

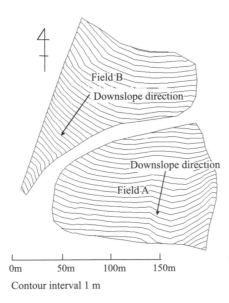

Fig. 1 Irregular shaped slope grassland

Fig. 2 Fertilizer application by fertilizing robot for steep slope use ("Kametarō")

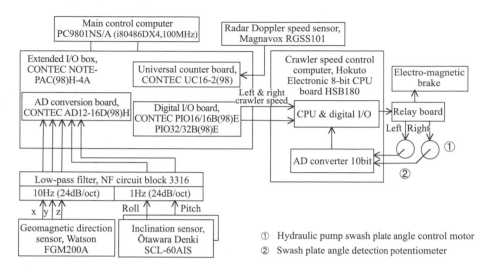

Fig. 3 System components of fertilizing robot for steep slope use

controlled by actuators such as electric motors and erector-magnetic valves. It has a navigation system to find its own position, a main control computer and a crawler control computer (Figure 3). The allowable deviation from the target travel path during unmanned operation was set at a maximum of 2.5 meters by reference to conventional operations by a skilled operator. In view of this permissible error, the navigation system is relatively low-cost, consisting of a **radar Doppler speed sensor** for speed detection, a **geomagnetic direction sensor**[1-4.2] GDS for vehicle heading direction detection, and a liquid level **inclination sensor** to compensate the GDS measurement. This system estimates the current position of the robot as a cumulative value from the operation start point based on speed and direction data. Unmanned operation is performed on the basis of the coordinates of the target travel path that are obtained from the previously surveyed data and entered in the main control computer. The robot is navigated by the software which corrects its course if a position deviation from the target travel path observed by the navigation system is 0.5 meters or more. The target travel path in an irregular shape and sloping field is set in consideration of the application width and hazard spot avoidance.

Operating on an approximately 10 degree slope at a speed of 0.5 m/s along Path 1 in Figure 4, the deviation of the actual travel path from the target travel path, that is the navigation error, was up to 2.5 meters and the difference between the start point and the end point was 4.6 meters. The navigation error was up to 2.5 meters and the difference between the start point and the end point was 4.9 meters in the case of Path 2. Both results were within the acceptable range for practical use.

The accuracy of **dead reckoning navigation** on sloping grassland is explained in detail in an article listed in the references.[45]

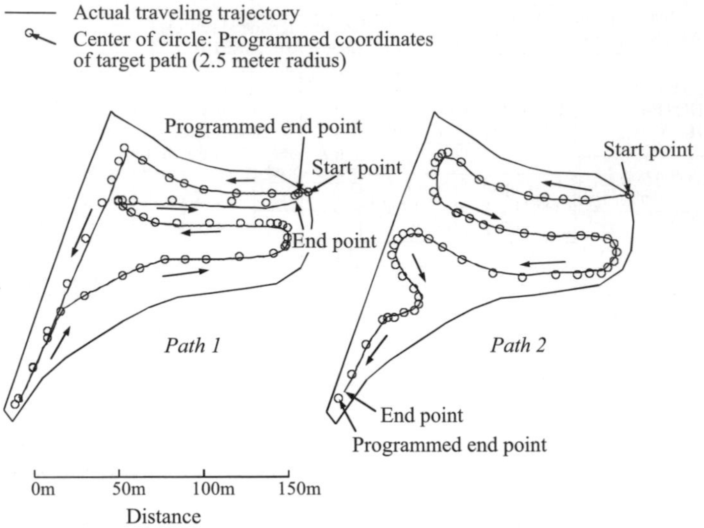

Fig. 4 *Trajectory of fertilizing robot on steep grassland (Field B in Figure 1)*

4.18 Monorail systems

A multipurpose monorail system was developed by a team led by BRAIN in 2001. It travels on monorail tracks laid in steep sloping orchards in the directions of inclination and contour to perform unmanned chemical spraying and transportation. There are basically two types: the round-about type and branch type track layouts.

Monorails are widely used for transporting products and farm materials up and down steeply sloping orchards. However, transporting in the contour direction often relies on human-powered carriers such as wheelbarrows, and chemical spraying for pest control is done using hoses from **power sprayers** manually pulled into the orchard by workers. These operations impose burdens on workers and are relatively inefficient. The **multipurpose monorail** system for steeply sloping orchards was developed at BRAIN in 2001 and includes a round-about type track layout and a branch type track layout.[46]

The round-about type can be used in sloping orchards with a gradient of up to 30° and consists of a main rail laid in the direction of slope inclination, an S-shaped rail running in the direction of contour lines, a main rail tractor, a main rail cargo carrier, an engine drive S-rail tractor and S-rail operation machines (chemical sprayer, rail tank car, cargo carrier and fertilizer applicator) (Figure 1). Products are loaded on the S-rail cargo carrier, transported by the S-rail tractor in the contour direction, manually transferred to the main rail cargo carrier at a point where the S-rail runs adjacent to the main line, and then transported up and down the slope by the main rail tractor. In chemical spraying, the S-rail tractor tows a tank car and the chemical sprayer with a pump, a blower, and so on (Figure 2). It is capable of both air injection spraying and

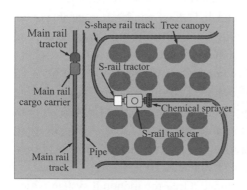

Fig. 1 Round-about type track layout

Fig. 2 Spraying by multipurpose monorail (round-about type)

299

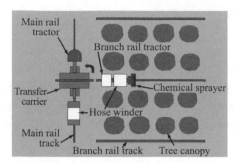

Fig. 3 Branch type track layout

airless spraying. The spray head can be rotated to either side and sprays chemicals to one side of the track on the outward journey and the other side on the return journey. Chemical spraying operation and spray head orientation control are done automatically. The chemical solution is supplied through a pipe laid along the main rail track and refilling is carried out at the storage tank at the end of the main line or a point of the S-rail adjacent to the main line.

The branch rail type layout is used in very steep sloping terrace-type orchards and consists of a main rail line in the direction of slope inclination, branch rail lines in the direction of contour lines, a main rail tractor, a main rail transfer carrier, an electric-powered branch rail tractor and branch rail operation machines (similar to those for the round-about type) that are towed by the branch rail tractor (Figure 3). Products are loaded on the branch rail cargo carrier and towed by the branch rail tractor in the contour direction. Both the branch rail tractor and the cargo carrier drive on to the transfer carrier on the main rail line and the transfer carrier is towed up and down the slope by the main rail tractor. In chemical spraying, the branch rail tractor tows the chemical sprayer car but the chemical solution is supplied to the sprayer car through a hose from a chemical tank and power sprayer located at the end of the main rail line. The chemical sprayer car is fitted with a blower, a spray head and a hose winder. It is capable of air injection spraying and airless spraying. It sprays with unwinding the spray hose. The branch rail tractor and chemical spray car can be operated by radio control.

The use of such multipurpose monorails can reduce the amount of chemicals used by about 50% of that for sprinkler spraying systems. The use of unmanned or remote control spraying can prevent chemical exposure of orchard workers and increase operating efficiency. In addition to labor saving in transportation and fertilizer application, the cargo carrier can be mounted with a branch shredder machine to process pruned branches or a power sprayer and tank to spray herbicides.

In 1993, NARC for Western Region developed a tree-top monorail system which ran on a rail track installed above tree canopies using support columns for the trellis-trained citrus growing system, and which is used for automatic crop transportation and chemicals spraying (Figures 4 & 5).[47] In harvesting by the tree-top monorail, a container carrier travels on the rail track above the trees and collects manually harvested fruits from a basket type conveyor running next to the container carrier. The container is carried out of the orchard on a monorail transportation vehicle. In spraying operation, the tractor tows a pest control machine fitted with a power sprayer, a blower, booms and a chemical tank and sprays the chemicals from both sides of the trees. Spraying operation can be controlled automatically at a rate of about 6 min/ 0.10 ha.

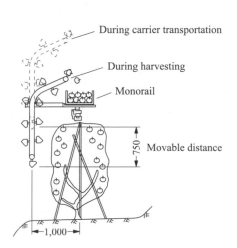

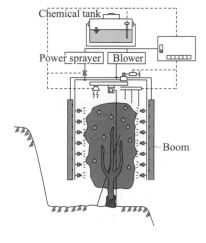

Fig. 4 Tree-top monorail (harvesting)
(NARC for Western Region)

Fig. 5 Tree-top monorail (spraying)

4.19 Gantry systems

Research on gantry systems in Japan was conducted at NARC and BRAIN during the 1970s. The implement fitted at the center of the beam performs various tasks while the traveling sections at each end of the gantry traverse prescribed areas without putting pressure on the soil, thus avoiding soil compaction problems.

A **gantry** is a machine which has traveling components at each end of a beam straddling the whole or part of a field. The implement attached to the beam performs tasks while the gantry traverses on rails or predetermined paths in the field. Gantries are either self-propelled type fitted with **crawlers** or wheels or an installation type which travels on rails laid at each end of the field.

While vehicles such as tractors compact the soil with repetitive treading pressure, gantries cause less soil compaction in the field since they traverse only on rails or prescribed paths. Their travel is stable, unaffected by soil conditions, which enhances operating efficiency due to wide operating range. In management operation, the cropping area can be increased as less land is set aside for traveling paths for farm vehicles. Installation type gantries can be easily automated since they travel on rails and can perform tasks with high accuracy due to precise positioning along rails. Self-propelled type gantries have been researched mainly in the U.S. and Europe and are already available on the market.

One example of an installation type gantry is the research gantry system developed at NARC in 1982 (Figure 1).[48] This system consists of four 10 m x 50 m

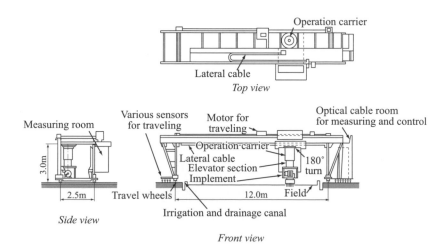

Fig. 1 Mechanism of gantry system (NARC)

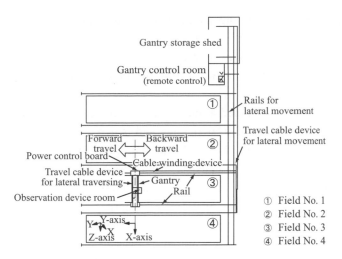

Gantry storage shed

Gantry control room
(remote control)

① Rails for lateral movement

Travel cable device for lateral movement

Forward travel ⟵⟶ Backward travel ②

Power control board

Cable-winding device

Travel cable device for lateral traversing

Gantry Rail ③

Observation device room

Y-axis
Y X
Z-axis X-axis ④

① Field No. 1
② Field No. 2
③ Field No. 3
④ Field No. 4

Fig. 2 Gantry system in field

fields, the control room and the storage shed (Figure 2). The weight of the gantry is 12 t (with a **combine** harvester attached) and the traveling speed is 0.1–0.4 m/s. The traveling device was driven by an electric motor. A composite cable consisting of a power transmission line and an optical cable is used for electricity supply and communication and the cable is wound and unwound according to gantry movement. The operating implement is attached to the frame (gantry section) and movable for 50 meters in the direction of rails (Y-axis), 12 meters in the lateral direction (X-axis) and 75 centimeters in the vertical direction (Z-axis). The gantry travels from the storage shed to the end of each field on travel rails, switches from traveling wheels to operating wheels, and performs its tasks as it traverses on rails laid along furrows (Figure 3).

Implements that can be attached to the gantry include **power tillers**, **fertilizing machines**, **seeders**, **transplanters**, **pest control machines** and harvesters. The

Fig. 3 Gantry system

Fig. 4 Gantry system for strawberry (2nd model) (Ehime Univ.)

gantry can achieve a positioning accuracy of ± 3 cm in the X direction, ± 4.5 cm in the Y direction and ± 1 cm in the Z direction by computer control, and can operate autonomously without a human operator.

BRAIN also researched gantries during the 1970s with the aim of automating farm operation.[49]

Ehime University developed a gantry for greenhouse use (Figure 4).[50(II-3.9)] It is used in soil cultivation of strawberries and is driven by an electric motor on square pipe rails laid in furrows, straddling two ridges (Figure 5). It can be fitted with work benches and a chemical sprayer and can be used for seated strawberry picking, tilling, ridging and pest control. It uses another carrier when transferring to another ridge. The traveling speed of the gantry is 0–70 cm/s. When it was trialed for strawberry picking operations, the efficiency was slightly lower than conventional operation but there were some marked workload reduction effects such as a 50 percent reduction in oxygen consumption by workers and significantly reduced burden on their muscles.

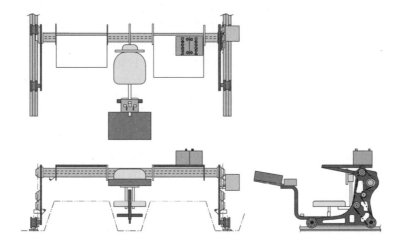

Fig. 5 Construction of gantry system for strawberry

4.20 Crawler type robot

(Movie)

> The crawler type robot was developed by a team led by the Hokkaido Industrial Research Institute in 2000. Since crawler type vehicles can operate on sloping ground where wheel type tractors cannot enter and their low ground contact pressure enables them to do farming work on soft ground and snow covered areas, the robotization of such vehicles is an important issue. Their steering system is different from that of wheel type vehicles and crawler speed control is an important issue.

The **crawler type robot** developed at the Hokkaido Industrial Research Institute is described below (Movie).[51–53 (I-4.3)] The machine consists of a **crawler tractor** and the fertilizer applicator with separate controllers. The two controllers are synchronized through interdependent control information communication by CAN network connection. Figure 1 shows the developed robot.

The base vehicle is an Ishikari Zōki **snow melting chemical applicator** US-D with 25 PS rubber **crawlers**. It consists of two sets of **HST**s (Hydro Static Transmissions) to control the speed of the left and right crawlers, auxiliary pumps to assist with steering and drive the implement, and a diesel engine to supply power to these components. The crawler speed varies according to the discharge rate from the HST pumps. Accordingly, the rotational output of the servo motor is controlled directly by the rotation position of the HST lever so that the speed of the left and right crawlers can be controlled independently of each other.

The mounting arrangement for the servo motor is shown on the right side of Figure 1 and the block diagram of the HST control system is shown in Figure 2. For the engine, the fuel pump on/off and the engine speed are controllable. Engine speed is controlled by one of eight engine throttle positions by an electric actuator.

Base machine

HST and servo motor

Fig. 1 Crawler type robot tractor and HST equipped with servo motor (Hokkaido Industrial Research Institute)

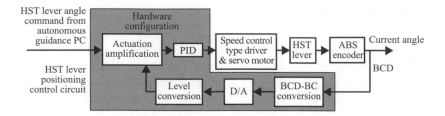

*Fig. 2 Control block diagram of HST for steering (Hokkaido
Industrial Research Institute)*

While emergency stop push buttons located at various places on the vehicle body
are the only safety devices at this stage, there are plans to fit additional devices
such as non-contact obstacle sensors in the future.

The implement is capable of various applications. It can apply snow melting
chemicals or fertilizers, adjusting quantities according to an application map.
The application quantity of snow melting chemicals or fertilizer in the hopper
is controlled by a variable opening gate valve installed immediately before the
applicator. An electric cylinder is used as the actuator for the variable opening
gate. Information about the status of the implement, vehicular information and
operational commands required for application are sent and received by the
micro computer board via CAN and the implement operates in response to such
information.

Steering is controlled by the two HSTs. The computation of navigation signals,
including lateral deviations and directional deviations, is performed in the same way
as that for wheeled vehicles. That is, they are calculated based on the **target path**,
the current position and the current direction. It uses an **RTK-GPS**[1-4.2] (Trimble
MS750) for positioning and a **fiber optic gyroscope** (FOG)[1-4.2] (Japan Aviation
Electronics Industry JD-108FD) for direction measurement.

It can perform **autonomous guidance** on sloping terrain by incorporating a
function to correct a positional error from vehicular inclination. The actual travel
tests on a ski slope at a gradient of around 20°, a snow-covered field and a flat road
surface have confirmed that the robot can generally travel with an accuracy of ± 5
cm. Figure 3 shows a recorded travel path of the crawler robot in the snow. Circles
show the target path and the broken line shows the actual travel path. It indicates that
the robot followed the target path accurately, including the curvilinear segments.

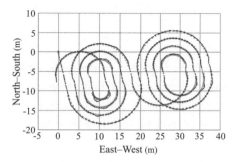

*Fig. 3 Test result of autonomous operation on snowfield (Hokkaido
Industrial Research Institute)*

4.21 Legged robots

Legged robots are attracting attention in recent years. Off-road ground surfaces such as farm land sometimes demand drivability, step-climbing ability and gradability, and the capabilities of multi-legged robots seem appealing. Walking type robots for forestry are undergoing testing at the moment but further development of these types of robots is highly anticipated.

The wheel type travel mechanisms that are used widely in transportation robots are suitable for traveling on level surfaces but they have difficulty traveling on uneven surfaces. Since **legged robots** have discrete ground contact points, they can travel under complex conditions by climbing over obstacles, straddling, avoiding holes and other bad footing areas and changing the height and attitude of the machine body.[1-4.3] Due to these advantages, legged robots are expected to find a wide range of applications such as building and construction, disaster rescue operations and land mine clearing.

Figure 1 shows the domestic utility quadruped **walking robot** BANRYU from TMSUK Co., Ltd. which came on the market in 2003. It uses three photo sensors fitted on its legs to detect height differences and climb over steps using one of three programmed walking patterns (0–5 cm, 5–10 cm and 10–15 cm). Since it is designed to travel on the floor inside the house, there is no need to consider rollover risks due to attitudinal change on soft ground, hence walking control of this robot is relatively easy.

This robot can receive commands for movement via a mobile phone and transmit moving images of its surroundings taken by its built-in camera. Its various sensors, including motion sensors (infrared) and odor sensors, are turned on when the occupant of the house goes out and the robot stands on its legs and sounds an alarm when these

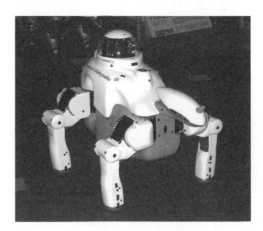

Fig. 1 Four-legged domestic utility robot (BANRYU) (TMSUK Co., Ltd.)

sensors detect their objects. At the same time, it rings the controller mobile phone and sends images of its surroundings.

Achieving the right cycle of walking motion, or a gait, is a challenge for multi-legged robots used on irregular or soft surfaces. Generating appropriate gait patterns is important for the harmonious and stable operation of all joints. It is also important that the robot has an ability to continue walking when it encounters irregular ground surfaces or unexpected breakdowns. There is a need for autonomous and heuristic gait patterns in order to deal with events such as unpredictable rollovers of the robot itself and excessive loads on its legs, for example.

In order to achieve this, there are attempts to develop automatic learning of gait patterns by trial and error, using sensors to monitor if the robot is meeting given objectives and trying to receive correct sensor signals more frequently. The reinforcement learning model can be used for this purpose. The generation of gait patterns suitable for walking on irregular ground surfaces based on the observation and modeling of the motions of insects is also in the research phase. This approach generally uses **force sensors** on the legs and attitude angle sensors on the body to measure and take account of subgrade bearing capacity and the attitude of the robot.

Figure 2 shows the quadruped wall walking robot NINJA developed at Tokyo Institute of Technology.[54] It is designed to perform hazardous operations at high places such as external wall inspection and painting for buildings and has suction cups at the end of each leg. Since NINJA has a motor configuration called 'interference drive', it can realize a greater power output when it performs vertical upward climbing attached to a wall. It can transfer itself from the floor to the wall, from the wall to the ceiling, and between two walls at a right angle, and climb up a wall at a speed of 13 cm/s.

Figure 3 shows the six-legged walking robot under development at Timberjack, Ltd. Although it is a manually operated robot, it represents a leading-edge technology in legged vehicles for outdoor use. While it is still in the development phase, it can travel

Fig. 2 Quadruped walking robot (NINJA) (Tokyo Institute of Technology)

Fig. 3 Six-legged walking robot for forestry (Timberjack, Ltd.)

on very steep slopes that are unsuited to wheel or **crawler** type vehicles. In forest land, wheel and crawler type vehicles disturb the ground surface and cause damage to the tree root area but such ground disturbance is minimal in the case of leg type vehicles. This robot detects the ground contact pressure of each leg, joint angles and the attitude angle of the body in selecting leg positions so that its six legs can support the body evenly. The leg type vehicle is highly drivable due to its capacity to perform diagonal and lateral movements and pivot turning as well as forward and backward movements. The operator of this robot can also change the height of the body or the leg lift height for each step.

Robot helicopters with autonomous flight capabilities have been researched and developed for various uses in natural disasters and environmental monitoring. Commercial models have been available since 2000. They have fewer limitations as to flying distance and altitude because their navigation does not rely on visual confirmation by the operator and they are expected to find applications in the monitoring of hazardous areas, the global environment and the regional environment.

Research on the automation and robotization of small unmanned **helicopters** has been underway both in Japan and overseas.[55] In Japan, corporation-based R & D activity is widespread. The **autonomous flight** system developed by Yamaha Motor Co., Ltd. is being used at disaster sites and for environmental monitoring. It is well known that Yamaha's autonomous helicopters equipped with this system demonstrated their effectiveness in monitoring the situations during the eruptions of Mount Usu in April 2000 and Miyake Island in February 2001. Overseas, research has been conducted mainly at universities and there have been reports on the application of technologies such as learning-based control laws, robust control theory and PID gain scheduling to flight control. The robotization of helicopters is a relatively new research field and there is no established dynamic model or control method as yet.[I-4.3]

While robot helicopters range from small model helicopter-based ones to **unmanned industrial helicopters**, autonomous or robotic helicopters are generally equipped with devices such as an inertial measurement unit (IMU), **geomagnetic direction sensors**[I-4.2] and **GPS**[I-4.1] and need position control, speed control and **attitude control**. They also require mission planning, including **path planning**, and a ground station for flight monitoring. Figure 1 shows Yamaha's autonomous helicopter. Figure 2 shows the system display with which the operator can

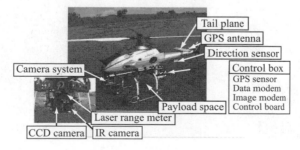

Fig. 1 Autonomous navigation type helicopter (Yamaha Motor Co., Ltd.)

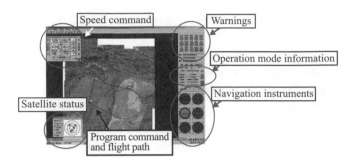

Fig. 2 Monitoring display for autonomous helicopter (Yamaha Motor Co., Ltd.)

monitor the flight path and status and the navigation mode as well as the status of GPS reception and the helicopter's instruments. In this helicopter, the system performs sequential computation of the target attitude angle, position and speed based on **target path** information and speed directives provided to hybridize the **feedforward control** system and the **feedback control** system and controls the helicopter's position, speed and attitude accordingly to guide the helicopter to the target path.

The robot helicopter for acrobatic flight has been researched at Massachusetts Institute of Technology (MIT) (see Figure 3) is also equipped with a GPS, an IMU and a geomagnetic direction sensor. It is controlled by the target yaw speed and the target transverse velocity calculated and transmitted to the helicopter by the ground station (Movie). It has an **extended Kalman filter**[1-4.3] to improve the precision of

Fig. 3 Robot helicopter system for acrobatic flight (MIT Aeronautics and Astronautics)

the position and attitude data with the addition of feedforward control based on rules devised from the skilled operator's navigation procedures. Flight monitoring and the computation of controlled variables based on the mission plan are performed on the ground station computer. The system can be switched over to normal radio control by the operator. In R & D of this type of robot, it is essential to develop a flight simulator to contain costs and maximize safety.

Figure 4 shows a simulator for the development of robot helicopters for acrobatic flight. The simulation system is based on modeling the dynamics of the helicopters using exactly the same servo motors and other components. It can facilitate efficient robotic development by providing desktop evaluation of control characteristics.

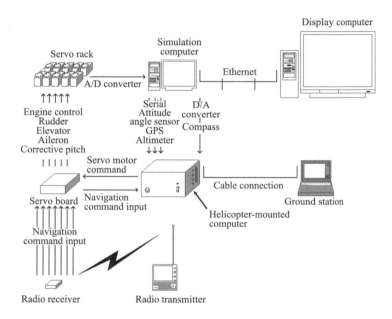

Fig. 4 Development simulator of robot helicopter for acrobatic flight
(MIT Aeronautics and Astronautics)

Master-slave system

> Systems for the coordinated operation of multiple farm transportation robots
> are promising as the next-generation robot systems. The system developed at
> Hokkaido University in 2003 adopts a coordinated operation method called
> master-slave system by which one robot tractor is designated as the leader
> (master) and the other robot tractor(s) the follower(s) (slave).

Using **vehicle following control** as an example, the master in the **master-slave** system
issues commands to the slave in order to maintain certain relative angles and relative
distances while traveling, constantly sending its own vehicular information to the slave.
The slave performs autonomous vehicular control based on the vehicular information
about the master and the relative angle and distance commands received (Figure 1).[1-4.7]
The establishment of a hierarchical relationship between robots makes the role of each
robot clearer and defines the construction of the system more precisely.[56]

The two **robot tractors** developed at Hokkaido University are based on Kubota
tractors GL320 (master) and MD77 (slave) (Movie). Both have been modified so that the
engine speed, **PTO** and steering can be controlled automatically. The slave is capable
of automatic control of its transmission and engine speed as well. Figure 2 shows the
configuration of the two robot tractors. Each robot is equipped with an **RTK-GPS**[1-4.2]
and a **fiber optic gyroscope** (FOG)[1-4.2] as **navigation sensors** and a wireless router
required for communication between the robot tractor.

The **wireless LAN** is simply a wireless version of an Ethernet cable. The wireless
LAN is less restricted physically in terms of its coverage since it uses radio waves
for communication and does not incur telecommunication charges. However, its
communication range is limited to the service area of the radio waves. This issue
can be overcome by setting up access points and establishing cable connection
between them.

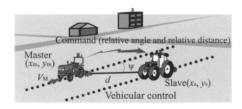

Fig. 1 Vehicle following control   Fig. 2 Configuration of master-slave system

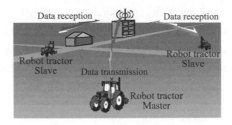

Fig. 3 Communication through access points

The test-manufactured master-slave system has been set up so that the robots can transmit and receive data through such access points (Figure 3). The existing protocols are designed to communicate general information and require many steps before exchanging information, hence they are not suitable for fast and accurate communication needed for measurement and control. For this reason, an original protocol has been developed for this system. With this protocol, the steps for the use of the **TCP** (Transmission Control Protocol) are taken at the time of connection between terminals and only the exchange of information occurs during measurement and control. The TCP is a protocol for data transmission control in process-to-process communication which is used when stable exchanges of highly reliable and accurate data are required. The TCP is effective in stable transmission and reception of accurate data when data communication is used for control of robot tractors.

Furthermore, the information being exchanged between the robots or the robots and the administrative host is divided into three types (command, status and request) (Figure 4). A command is transmitted by the master to the slave and contains the minimum information required for each action. A request is sent by the master to the slave when needed and asks the slave to provide information about the current state of operation. A status is sent by the master to the slave at regular intervals and informs the slave of its current position, speed and direction. The master also provides information about the current operation status when requested by the slave.[57]

The slave uses a **2D laser range finder** as a device to recognize and follow the master.[58] For the slave to acquire the forward direction and relative distance of the master, a template matching method has been adopted. The slave automatically acquires its relative angle and distance to the master by using the master's contour acquired by the 2D laser range finder and the contour template measured in advance. In other words, this system does not need position data transmission from the master. Sliding mode control is used as an algorithm for robot following control. Sliding mode control achieves desired dynamics by constraining the state of the system to a predetermined sliding plane in the state space according to the sliding conditions. It is often used for nonlinear control objects due to its superior robustness against modeling errors and disturbances. The slave executes follow-up traveling control via independent sliding mode control systems for speed and steering.[56]

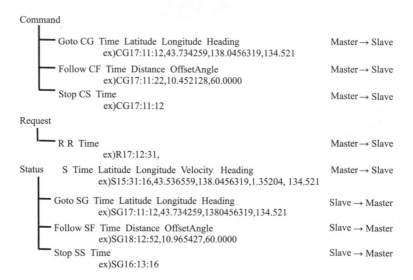

Fig. 4 List of data formats

4.24 Tele-operated robots (Movies)

> Tele-operation is a technology to control vehicles remotely which saves labor. However, there are still many issues to be resolved such as the development of an interface which allows humans to make optimal decisions. Vehicles must also have an autonomous function to navigate according to highly complex instructions. There are also issues of the best method of communication in outdoor environments.

Technology to operate vehicles by remote control (**tele-operation**) is being researched for applications where robots are expected to operate in places inaccessible to humans such as disaster sites, remote places such as the moon and planets, or extremely large- or small-scale environments. Cooperation with humans who are capable of judging the situation on a case-by-case basis in tele-operation enhances the intelligence of robots. In particular, its advantages are maximized when it is used in places where operating environments are less than optimal for robots. In other words, the technological concept of tele-operation is different from that of autonomous robots that are currently under research for full automation. In this section, tele-operation systems used for ground vehicles and the tele-operated **industrial unmanned helicopters** that are already on the market are described.[1-4.7]

The main issues for tele-operation of vehicles (Movies 1 & 2) include the application of X-by-wire technology to the hardware, the development of communication systems such as **wireless LAN**, and the user interface between humans and robots (Figure 1). X-by-wire is a system to perform vehicle control operations such as steering, gear change and braking using electric signals. Efficient system development is impossible unless the vehicle is equipped with this system because conversion and test-manufacture of the vehicle's main body cost time and money. It is also important that the robot is able to gather enough information for

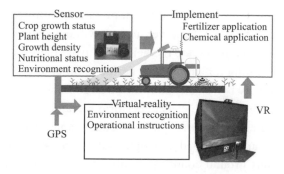

Fig. 1 Tele-operation by virtual reality system

Fig. 2 Industrial unmanned
helicopter

Fig. 3 Monitoring system for
helicopter flight

decision making by a human and communicate it to the ultimate decision maker, and that its control procedure is not burdensome to the operator. Any system which leaves all decision making to a human is obviously unacceptable. The robot vehicle and the human must cooperate and share the responsibility between them appropriately. For example, it is desirable that the vehicle can autonomously decide on matters such as the recognition and avoidance of mobile obstacles. The **master-slave** type tele-operation system under which simple operation procedures performed by the human are followed exactly by the machine is not applicable to vehicles for outdoor operation. It is essential to equip the vehicle with autonomous functions. Visual information is generally important for driving control and **stereo vision**[1-2,4] is an effective sensor for that purpose.

Industrial-use unmanned helicopters by tele-operation (Figure 2) are widely used in farming. They are mainly used in pest control operations and becoming popular among large-scale farms and contractors. A helicopter can spray chemicals on 2 hectares of land in a single 20-minute flight, which means that it can cover 20 to 30 hectares by operating for 5 hours a day. It has been reported that the helicopter can achieve the same level of application accuracy as conventional ground application methods because the helicopter's downwash disperses the chemical dust, which covers the vegetation evenly.[59]

In this way, industrial unmanned helicopters offer improved efficiency in chemical spraying operations. However, it goes without saying that the operator must have special skills to operate the helicopter from the ground. The operator needs to undergo training for a prolonged period in order to be able to control the helicopter appropriately by visually confirming its position and attitude regardless of its orientation. The operator needs to perform elevator control, rudder control, aileron control and engine control simultaneously and, at least at present, the operator cannot fly an unmanned helicopter safely and steadily without having the ability to convert ground coordinates into the helicopter's local coordinates instantaneously.[I-4.3] The **operations support system** is important in this respect and a flight monitoring system is a desirable function.

A flight monitoring system (Figure 3) was developed at Hokkaido University for improved efficiency and accuracy of helicopter-based **remote sensing**. It transmits data about the helicopter's attitude such as roll direction, pitch direction and **GPS**[I-4.1] position to the ground station via spread spectrum wireless communication. The monitoring system can achieve efficient and accurate sensing by blotting out the areas the helicopter has already worked on using a built-in **GIS**.[60] The system is most useful when the helicopter is flying at high altitude or far away from the operator since the operator can confirm the helicopter's attitude as well as its position on the display. It also has an image transfer device which allows the operator outside of the field to monitor what is happening inside the field. This system is also useful in minimizing waste in the conventional chemical spraying operation. Again, the important issue here is the interface between the operator and the unmanned industrial helicopter.

Closing remarks

Agriculture has sustained the survival and prosperity of human kind from the dawn of agrarian civilization, said to be in around 9000 B.C., to this day. The invention of ironware, steam engines and internal combustion engines by human ingenuity has resulted in a dramatic increase in food production. The continual growth in the number of people that can be fed by one producer using internal combustion engines is the basis for the advancement of today's secondary and tertiary industries.

The history of agriculture has been all about increasing production by increasing productivity. However, there are signs that this several thousand year old production strategy is now reaching its limit. The negative outcomes of our single-minded pursuit of higher productivity are manifesting as soil compaction by very large tractors and other farm machinery, environmental pollution by excessive use of agricultural chemicals and fertilizers, and concerns about food safety; among other things. It can be argued that in order to solve the food-related problems currently facing human kind, we are in need of production support in the form of not only powerful tractors and other machinery but also more sophisticated tools that can perform tasks with human-like finesse. This is the need that robots are expected to serve. Since the number of people who have to be supported by each food producer will continue to grow, the robot as a substitute for human workers involved in food production will be an indispensable part of the infrastructure for human survival in the future.

I believe that the readers of this book are now aware that the industrial, government and academic sectors are energetically researching the automation of, and robots for, food production operations. The first volume of this book, *Agri-Robot (I)*, deals with 'fundamentals and theory'. The second volume, *Agri-Robot (II)* describes 'mechanisms and practice' developed from the 'fundamentals and theory' and explains the already developed robots and automated systems. It describes a wide range of machines from those which are already in practical use to those at the development stage. However, understanding the current level of technological development is an essential part of the learning process in grasping the current state of robotics, discovering the seeds of robotic research and promoting the fruits of such activities. From this viewpoint, the authors of *Agri-Robot (II)* endeavored to explain these research results as concisely and plainly as possible. Moving images of robots in operation are included in the accompanying CD-ROM. Observing the actions of robots is most important for the understanding of the robots. These video clips have been collected by the authors through their friends and acquaintances all over the world. This is the first agricultural robot book in the world to include a CD-ROM containing moving images. It is a great pleasure for myself as one of the authors to have achieved this. Finally, I would like to close my remarks by expressing my sincere appreciation for the great help provided to us by Corona Publishing Co., Ltd. from the planning stage to the publication of this book.

April 2006
Noboru Noguchi

References

Chapter 1

1   BRAIN (Seibutsukei Tokutei Sangyō Gijutsu Kenkyū Suishin Kikō [Bio-oriented Technology Research Advancement Institution]) (2003), *21-seikigata nōgyō kikai tō kinkyū kaihatsu jigyō no seika ni tsuite (heisei 10 nendo–heisei 14 nendo)* (Report on the outcomes of the Urgent Development Project for Agricultural Machinery in the 21st Century (1998–2002)); and, *Nōgyō kikai tō kinkyū kaihatsu jigyō no seika ni tsuite (heisei 5 nendo–heisei 9 nendo)* (Report on the outcomes of the Urgent Development Project for Agricultural Machinery (1993–1997)).

2   Deere & Company website (as at 1 November 2005), http://www.deere.com/de_DE/news/2004/autotrac.html.

3   Shibusawa, S. (2003), 'On-line real time soil sensor', *IEEE/ASME International Conference on Advanced Intelligent Mechatronics*, CD-ROM.

4   Shibusawa, S. (2002), 'Community-based precision farming for small farm agriculture', *Proceedings of 6th International Conference on Precision Agriculture*, CD-ROM.

5   Kondō, Naoshi et al. (1998), 'Kiku no sashiki sagyō no jidōka ni kansuru kisoteki kenkyū (dai 1 pō) (Basic research on automation of chrysanthemum cutting sticking operation (part 1))', *Nōgyō Kikai Gakkaishi* (Journal of the Japanese Society of Agriculture Machinery), 60(2): 67–74.

6   Kondō, Naoshi et al. (1998), 'Kiku no sashiki sagyō no jidōka ni kansuru kisoteki kenkyū (dai 2 hō) (Basic research on automation of chrysanthemum cutting sticking operation (part 2))', *Nōgyō Kikai Gakkaishi* (Journal of the Japanese Society of Agriculture Machinery), 60(3): 63–70.

7   Monta, Mitsuji et al. (1998), 'Kiku no sashiki sagyō no jidōka ni kansuru kisoteki kenkyū (dai 3 pō) (Basic research on automation of chrysanthemum cutting sticking operation (part 3))', *Nōgyō Kikai Gakkaishi* (Journal of the Japanese Society of Agriculture Machinery), 60(4): 37–44.

8   Monta, Mitsuji et al. (1998), 'Kiku no sashiki sagyō no jidōka ni kansuru kisoteki kenkyū (dai 4 pō) (Basic research on automation of chrysanthemum cutting sticking operation (part 4))', *Nōgyō Kikai Gakkaishi* (Journal of the Japanese Society of Agricultural Machinery), 60(5): 37–43.

9   Kondō, Naoshi et al. (1999), 'Kiku no sashiki robotto no tameno sashiho bunri kyōkyū shisutemu no kaihatsu (Development of the cuttings separation and feeding system for the chrysanthemum cutting sticking robot)', *Nōgyō Kikai Gakkaishi* (Journal of the Japanese Society of Agriculture Machinery), 61(5): 109–116.

10 Watake, Hiroaki (1992), '7 sho jidōka (Chapter 7 automation)', *Haiteku nōgyō handobukku* (High-tech agriculture handbook), Tokai Daigaku Shuppankai, pp. 115–135.

11 Marusan Seiyaku K.K. website (as at 1 November 2005), http://www.marusans. co.jp/top/outline/jigyou/ms-x7/.

12 Kawamura, Noboru et al. (1984), 'Nōgyō robotto no kenkyū (dai 1 pō) (Agricultural robotics research (part 1))', *Nōgyō Kikai Gakkaishi* (Journal of the Japanese Society of Agriculture Machinery), 46(3): 353–358.

13 Kondo, N. and K. D. Ting (1998), *Robotics for Bioproduction Systems*: ASAE.

14 Ishii, Tōru et al. (2003), 'Rakuyōkei kajitsu senbetsu robotto (dai 1 pō) (Deciduous fruit grading robot (part 1))', *Nōgyō Kikai Gakkaishi* (Journal of the Japanese Society of Agriculture Machinery), 65(6): 163–172.

15 Ishii, Tōru et al. (2003), 'Rakuyōkei kajitsu senbetsu robotto (dai 2 hō) (Deciduous fruit grading robot (part 2))', *Nōgyō Kikai Gakkaishi* (Journal of the Japanese Society of Agriculture Machinery), 65(6): 173–183.

16 Ishii, Tōru et al. (2002), 'Shironegi jidō senbetsu shisutemu (Automatic leek grading system)', *Nōgyō kankyō kōgaku kanren 4 gakkai 2002 nendo gōdō taikai kōen yōshi, Nihon shokubutsu kōjō gakkai gakkaishō jushō kinen kōen yōshi* (Abstracts of papers of the joint conference on environmental engineering in agriculture 2002, abstracts of Japanese Society of High Technology in Agriculture Award Lectures), pp. 407–411.

17 Kondō, Naoshi (2003), 'Rōtarī baketto wo mochiita nasu kajitsu no senka shisutemu (Eggplant fruit grading system using rotary bucket)', *Nihon kikai gakkai robotikusu mechatoronikusu kōenkai 03 kōen ronbunshū* (The Japan Society of Mechanical Engineers Robotics and Mechatronics Conference 03 transactions), CD-ROM.

18 Ollice, M. and A. Stentz (1996), 'First results in crop line tracking', *Proceedings of IEEE Conference on Robotics and Automation*, pp. 951–961.

19 Usman, A. et al. (1998), 'Weed detection in lawn field based on gray-scale uniformity', *Environment Control in Biology*, 36(4): 227–237.

20 Usman, A. et al. (1999), 'Weed detection in lawn field using machine vision – utilization of textural features in segmented area', *Journal of the Japanese Society of Agriculture Machinery*, 61(2): 61–69.

21 Usman, A. et al. (1999), 'Weed center detection in lawn field using morphological image processing', *Journal of Society of High Technology in Agriculture*, 11(2): 127–135.

22 Trevelyan, J. P. (1992), *Robots for Shearing Sheep: Shear Magic*, London: Oxford University Press.

23 Shibusawa, S. (2003), 'A role of bio-production robots in precision farming Japan model', *IEEE/ASME International Conference on Advanced Intelligent Mechatronics*, CD-ROM.

24 Kondo, N. and S. Shibusawa (2003), 'Role of bio-production robot in precision farming', *Programme Book of the Joint Conference of ECPA-ECPLF*, pp. 513–514.

25 Shibusawa, Sakae (2003), 'Seimitsu nōgyō no kenkyū kōzō to tenbō (Research structure and outlook for precision farming)', *Nōgyō Jōhō Kenkyū* (Journal of Japanese Society of Agricultural Informatics), 12(4): 259–274.

26 Qiao, J. et al. (2004), 'Mobile fruit grading robot (Part 1), *Journal of the Japanese Society of Agriculture Machinery*, 66(2): 113–122.

27 Nōgyō Kikai Gakkai (ed.) (1996), *Seibutsu seisan kikai handobukku* (Handbook of bio-production machines), Tokyo: Corona Publishing, pp. 450–459.

28 Nishimura, Hiroshi et al. (2003), 'Kahen shihiki no kaihatsu (dai 1 pō) (Development of variable fertilizer applicator (part 1))', *Dai 62 kai Nōgyō Kikai Gakkai nenji taikai kōen yōshi* (Abstracts of papers of the 62nd annual conference of the Japanese Society of Agriculture Machinery), pp. 261–262.

29 Chōsa, Tadashi et al. (2003), 'Ryūjōbutsu sanpuki no mappu bēsu kahen seigyo shisutemu (Map-based variable control system for granule applicator)', *Nōgyō Kikai Gakkaishi* (Journal of the Japanese Society of Agriculture Machinery), 65(3): 128–135.

30 Nihon Nōgyō Kikaika Kyōkai (Japan Agricultural Mechanization Association) (2000), *Saishin nōgyō kikai shisetsu gaidobukku* (New edition handbook of agricultural machinery and plant), pp. 50–58.

31 Itō, Toshio (2003), 'Mugi daizu no fukōki saibai gijutsu (No-till farming technology for wheat and soy)', *Green Report* (monthly JA magazine), No. 412, pp. 2–4.

32 Nihon Nōgyō Kikaika Kyōkai website (as at 1 November 2005), http://nitinoki.or.jp/bloc1/kouhou/aug/01.htm#SEC4.

33 Konishi, Tatsuya (1997), 'Atarashii taue no gijutsu (New rice transplanting technology)', *Nōgyō Kikai Gakkaishi* (Journal of the Japanese Society of Agriculture Machinery), 59(4): 123–127.

34 Seike, Masanori (1997), 'Taueki no sōkō sōchi (Travel device for rice transplanter)', *Nōgyō Kikai Gakkaishi* (Journal of the Japanese Society of Agriculture Machinery), 59(3): 145–146.

35 Nōgyō Kikai Gakkai (ed.) (1996), *Seibutsu seisan kikai handobukku* (Handbook of bio-production machines), Tokyo: Corona Publishing, pp. 511–527.

36 BRAIN (Seibutsukei Tokutei Sangyō Gijutsu Kenkyū Suishin Kikō) (ed.) (2001), 'Kōseido suidenyō josōki no kaihatsu (Development of high precision rice paddy weeding machine)', *Heisei 12 nendo jigyō hōkoku, Seiken Kikō* (Annual report of the Bio-oriented Technology Research Advancement Institute for 2000), pp. 78–79.

37 BRAIN (Seibutsukei Tokutei Sangyō Gijutsu Kenkyū Suishin Kikō) (ed.) (1998), 'Kōseido suidenyō josōki no kaihatsu (Development of high precision rice paddy

weeding machine)', *Heisei 9 nendo jigyō hōkoku, Seiken Kikō* (Annual report of the Bio-oriented Technology Research Advancement Institute for 1997), pp. 118–119.

38 Zhang, Shu-huai et al. (2002), 'Konsairui yasai no mabiki sagyō no jidōka ni kansuru kenkyū (dai 2 hō) (Research on automation of thinning operation for root vegetables (part 2))', *Nōgyō Kikai Gakkaishi* (Journal of the Japanese Society of Agriculture Machinery), 64(2): 71–77.

39 Kawamura, Noboru et al. (1991), *Shinban nōsagyō kikaigaku* (New edition agricultural machinery), Tokyo: Buneidō Publishing, pp. 144–146.

40 Yanmar Co., Ltd. website (as at 1 November 2005), http://www.yanmar.co.jp/index/htm.

41 Kawamura, Noboru et al. (1991), *Shinban nōsagyō kikaigaku* (New edition agricultural machinery), Tokyo: Buneidō Publishing, pp. 198–207.

42 Kanemitsu, Mikio et al. (1993), 'Hakusai shūkakuki no kaihatsu kenkyū (dai 2 hō) (Research and development of Chinese cabbage harvester (part 2))', *Nōgyō Kikai Gakkaishi* (Journal of the Japanese Society of Agriculture Machinery), 55(6): 121–128.

43 Nōgyō Kikai Gakkai (ed.) (1996), *Seibutsu seisan kikai handobukku* (Handbook of bio-production machines), Tokyo: Corona Publishing, pp. 789–877.

44 Njoroge, J. et al (2002), 'Automated fruit grading system using image processing', *Proceedings of SICE Annual Conference 2002 in Osaka* , CD-ROM.

45 BRAIN (Seibutsukei Tokutei Sangyō Gijutsu Kenkyū Suishin Kikō) (2003), *21-seiki gata nōgyō kikai tō kinkyū kaihatsu jigyō no seika ni tsuite (heisei 10 nendo–heisei 14 nendo)* (Report on the outcomes of the Urgent Development Project for Agricultural Machinery in the 21st century (1998–2002)), *Nōgyō kikai tō kinkyū kaihatsu jigyō no seika ni tsuite (heisei 5 nendo–heisei 9 nendo)* (Report on the outcomes of the Urgent Development Project for Agricultural Machinery (1993–1997)), pp. 1–3.

46 Nihon Nōgyō Kikaika Kyōkai (Japan Agricultural Mechanization Association) (2000), *Saishin nōgyō kikai shisetsu gaidobukku* (New edition handbook of agricultural machinery and plant), pp. 199–229.

47 Shitō, Hirokatsu (2003), 'Aogari tōmorokoshi shūkaku to kikai (Soiling corn harvesting and machinery)', *Nōgyō Kikai Gakkaishi* (Journal of the Japanese Society of Agriculture Machinery), 65(6): 4–8.

48 Nōgyō Kikai Gakkai (ed.) (1996), *Seibutsu seisan kikai handobukku* (Handbook of bio-production machines), Tokyo: Corona Publishing, pp. 360-370.

49 Kanemitsu, Mikio et al. (2003), 'Keishachi kaju yō tamokuteki monorēru no kaihatsu (dai 2 hō) (Development of multipurpose monorail for fruits growing on steep orchards (part 2)), *Dai 62 kai Nōgyō Kikai Gakkai nenji taikai kōen yōshi* (Abstracts of papers of the 62nd annual conference of the Japanese Society of Agriculture Machinery), pp. 83–84.

Chapter 2

1 Shibusawa, Sakae and Shinichi Hirako (2001), 'Seimitsu nōhō no tameno riarutaimu dochū hikari sensā (Real-time light sensor for precision farming)', *Bunkō kenkyū* (Journal of the Spectroscopical Society of Japan), 50(6): 251–260.

2 Shibusawa, S. et al. (2002), 'Soil mapping strategy using real-time soil spectrophotometer', *Proceedings of the 6th International Conference on Precision Agriculture*, CD-ROM.

3 Roy, S. K. et al. (2004), 'Characterization of fertilizer and manure stressed soil images with textural analysis using real-time soil spectrophotometer', *Proceedings of the 8th International Conference on Precision Agriculture*, CD-ROM.

4 Christy, D. D. et al. (2003), 'An on-the-go spectral reflectance sensor for soil', *ASAE Paper No. 031044,* St. Joseph, Michigan: American Society of Agricultural and Biological Engineers.

5 Mouazen, A. M. et al. (2005), 'Towards development of on-line soil moisture content sensor using a fibre-type NIR spectrophotometer', *Soil & Tillage Research*, 80, pp. 171–183.

6 Kondo, N. et al. (1996), 'Visual sensing algorithm for chrysanthemum cutting sticking robot system', *Acta Horticulturae*, 440, pp. 383–388.

7 Kondo, N. et al. (1997), 'Cutting providing system and vision algorithm for robotic chrysanthemum cutting sticking system', *Preprints of the International Workshop on Robotics and Automated Machinery for Bioproductions*, pp. 7–12.

8 Kondō, Naoshi et al. (1998), 'Kiku no sashiki sagyō no jidōka ni kansuru kiso kenkyū (dai 1 pō) (Basic research on automation of chrysanthemum cutting sticking operation (part 1))', *Nōgyō Kikai Gakkaishi* (Journal of the Japanese Society of Agriculture Machinery), 60(2): 67–74.

9 Kondō, Naoshi et al. (1998), 'Kiku no sashiki sagyō no jidōka ni kansuru kiso kenkyū (dai 2 hō) (Basic research on automation of chrysanthemum cutting sticking operation (part 2))', *Nōgyō Kikai Gakkaishi* (Journal of the Japanese Society of Agriculture Machinery), 60(3): 63–70.

10 Watake, Hiroaki (1991), 'Nae zōshoku yō robotto (Seedling propagation robot)', *Shokubutsu saibō kōgaku* (Plant cell engineering), 3(1): 73–77.

11 Simonton, W. (1990), 'Automatic geranium stock processing in a robotic workcell', *Transactions of ASAE*, 33(6): 2074–2080.

12 Simonton, W. and J. Peace (1990), 'Automatic plant feature identification of geranium cuttings using machine vision', *Transactions of ASAE*, 33(6): 2067–2073.

13 Dohi, Makoto et al. (1993), 'Yasai yō takinō robotto no kenkyū (dai 1 pō) (Research on multifunctional robot for vegetable production (part 1))', *Nōgyō Kikai Gakkaishi* (Journal of the Japanese Society of Agriculture Machinery), 55(6): 77–84.

14 SICK, Inc., LMS200 laser scanner system technical information.

15 Monta, Mitsuji et al. (2002), 'Ningen kyōchō gata nōgyō yō robotto no gekai

senshingu shisutemu (dai 1 pō) (External sensing systems for human cooperative agricultural robots (part 1))', *Shokubutsu kōjō gakkaishi* (Journal of the Society of High Technology in Agriculture), 14(1): 49–55.

16 Monta, Mitsuji et al. (2002), 'Ningen kyōchō gata nōgyō yō robotto no gekai senshingu shisutemu (dai 2 hō) (External sensing systems for human cooperative agricultural robots (part 2))', *Shokubutsu kōjō gakkaishi* (Journal of the Society of High Technology in Agriculture), 14(2): 104–111.

17 Gotou, K. et al. (2003), '3-D vision system of tomato production robot', *Proceedings of the IEEE/ASME International Conference on Advanced Intelligent Mechatronics*, pp. 1210–1215.

18 Subrata, I. D. M. et al. (1997), '3-D vision system for tomato harvesting robot', *JARQ*, 31(4): 257–264.

19 Kondo, N. et al. (1996), 'Visual feedback guided robotic cherry tomato harvesting', *Transactions of the ASAE*, 39(6): 2,331–2,338.

20 Shibano, Yasunori et al. (1996), 'Shikaku fīdobakku hō wo riyō shita minitomato shūkaku robotto no seigyo hōhō (Control system for cherry tomato harvesting robot using visual feedback)', *Nōgyō Kikai Gakkai Kansai Shibu hō* (Report of the Japanese Society of Agriculture Machinery Kansai Branch), 80, pp. 47–48.

21 Kondō, Naoshi et al. (1999), 'Ichigo shūkaku robotto no tameno shikaku arugorizumu (Vision algorithm for strawberry harvesting robot)', *Nōgyō Kikai Gakkai Kansai Shibu hō* (Report of the Japanese Society of Agriculture Machinery Kansai Branch), 86, pp. 77–78.

22 Hayashi, Shigehiko et al. (2003), 'V-jigata seishisareta nasu wo taishō to shita robotto shūkaku shisutemu (dai 1 pō) (Robot harvesting system for eggplants trained on a V-shape trellis (part 1))', *Shokubutsu kōjō gakkaishi* (Journal of the Society of High Technology in Agriculture), 15(4): 205–210.

23 Hayashi, Shigehiko et al. (2003), 'V-jigata seishisareta nasu wo taishō to shita robotto shūkaku shisutemu (dai 2 hō) (Robot harvesting system for eggplants trained on a V-shape trellis (part 2))', *Shokubutsu kōjō gakkaishi* (Journal of the Society of High Technology in Agriculture), 15(4): 211–216.

24 Kondō, Naoshi and Shunzō Endō (1988), 'Kajitsu ninshiki yō shikaku sensa no kenkyū (dai 3 pō) (Research on fruit recognizing visual sensor (part 3))', *Nōgyō Kikai Gakkaishi* (Journal of the Japanese Society of Agriculture Machinery), 50(6): 83–89.

25 Arima, S. and N. Kondo (1999), 'Cucumber harvesting robot and plant training system', *Journal of Robotics and Mechatronics*, 11(3): 208–212.

26 Karahashi, Motomu and Itō Shigeaki (1983), 'Kekkyū yasai shūkakuki no kaihatsu kenkyū (Research and development of harvester for head-forming vegetables)', *Nōgyō Kikai Gakkaishi* (Journal of the Japanese Society of Agriculture Machinery), 45(1): 71–72.

27 Murakami, Noriyuki et al. (1997), 'Gazō ni yoru kyabetsu ninshiki gijutsu no kaihatsu (Development of image-based cabbage recognition technology)', *Nōgyō Kikai Gakkaishi* (Journal of the Japanese Society of Agriculture Machinery), 59(2): 65–72.

28 Satō, Tadatoshi et al. (2001), 'Sanjigen shikaku sensa ni yoru kyabetsu kekkyū no keijō keisoku to shūkaku tekiki hantei (Shape measurement and harvesting stage determination by 3-D vision sensor for cabbage head)', *Nōgyō Kikai Gakkaishi* (Journal of the Japanese Society of Agriculture Machinery), 63(1): 87–92.

29 Njoroge, J. et al. (2002), 'Automated fruit grading system using image processing', *Proceedings of SICE Annual Conference 2002 in Osaka*, CD-ROM.

30 Ishii, Tōru et al. (2003), 'Rakuyōkei kajitsu senbetsu robotto (dai 1 pō) (Deciduous fruit grading robot (part 1))', *Nōgyō Kikai Gakkaishi* (Journal of the Japanese Society of Agriculture Machinery), 65(6): 163–172.

31 Ishii, Tōru et al. (2003), 'Rakuyōkei kajitsu senbetsu robotto (dai 2 hō) (Deciduous fruit grading robot (part 2))', *Nōgyō Kikai Gakkaishi* (Journal of the Japanese Society of Agriculture Machinery), 65(6): 173–183.

32 Ogawa, Y. et al. (2003), 'Inside quality evaluation of fruit by X-ray image', *Proceedings of the 2003 IEEE/ASME International Conference on Advanced Intelligent Mechatronics*, CD-ROM, pp. 1360–1366.

33 Kondo, N. et al. (2004), 'Eggplant grading machine by use of rotary trays', *Proceedings of Automation Technology for Off-Road Equipment*, St. Joseph, Michigan: American Society of Agricultural and Biological Engineers, pp. 394–398.

34 Ninomiya, K. et al. (2004), 'Machine vision systems of eggplant grading system', *Proceedings of Automation Technology for Off-Road Equipment*, St. Joseph, Michigan: American Society of Agricultural and Biological Engineers, pp. 399–404.

35 Itō, Takeshi (1998), 'Kyūri senbetsuki (Cucumber grading machine)', *Keisoku to seigyo* (SICE journal of control, measurement and system integration), 37(2): 103–104.

36 Qiao, J. et al. (2004), 'Mobile fruit grading robot (part 1)', *Journal of the Japanese Society of Agriculture Machinery*, 66(2): 113–122.

37 Qiao, J. et al. (2005), 'Mapping yield and quality using the mobile fruit grading robot', *Biosystems Engineering*, 90(2): 135–142.

38 Ishii, Tōru et al. (2002), 'Shironegi jidō senbetsu shisutemu (Automatic leek grading system)', *Nōgyō kankyō kōgaku kanren 4 gakkai 2002 nendo gōdō taikai kōen yōshi,* (Abstracts of papers of the joint conference on environmental engineering in agriculture 2002), pp. 408–411.

39 BRAIN (Seibutsukei Tokutei Sangyō Gijutsu Kenkyū Suishin Kikō) (2003), *21-seiki-gata nōgyō kikai tō kinkyū kaihatsu jigyō heisei 14 nendo kaihatsuki no gaiyō oyobi seisekisho* (An overview and performance assessment of the machines developed in

2002 under the Urgent Development Project for Agricultural Machinery in the 21st Century), pp. 25–26.

40 Ikeda, Yoshirō et al. (1996), 'Gazō shori ni yoru meron no gaikan hantei (Visual testing of melon by image processing)', *Nōgyō Kikai Gakkai Kansai Shibu hō* (Report of the Japanese Society of Agriculture Machinery Kansai Branch), 80, pp. 97–98.

41 Kawakami, Shōtarō et al. (1996), 'Netting patterns of melons', *Nōgyō Kikai Gakkai-shi* (Journal of the Japanese Society of Agriculture Machinery), 58(1): 17–23.

42 Sugiyama, Junichi (2003), *Shokuhin no hihakai keisoku handobukku* (Handbook for nondestructive measurement of food), Tokyo: Science Forum, pp. 245–251.

43 Hashimoto, Yasuo and Kazuyoshi Saitō (1998), 'NMR wo mochiita suika naibu hinshitsu kensa sōchi ni tsuite (Internal quality testing device for watermelon using NMR))', *Keisoku to seigyo* (SICE journal of control, measurement and system integration), 37(2): 108–110.

44 Katō, Kōrō (2001), 'Inpīdansu hō to mitsudo hō ni yoru suika no tōdo nikushitsu sokutei – tokusei hikaku to kōgakuteki hōhō tono fukugōka (Measurement of sugar content and texture of watermelon by impedance method and density method – comparison of characteristics and combining with optical methods)', *Dai 60 kai Nōgyō Kikai Gakkai nenji taikai kōen yōshi* (Abstracts of papers of the 60th annual conference of the Japanese Society of Agriculture Machinery), pp. 75–76.

45 Katō, Kōrō (1999), 'Suika no mitsudo senka ni kansuru kenkyū – Matsumoto hairando nōkyō suika shūshukka shisetsu (Research on watermelon density grading – Matsumono highlands agricultural cooperative watermelon handling facility)', *Dai 58 kai Nōgyō Kikai Gakkai nenji taikai kōen yōshi* (Abstracts of papers of the 58th annual conference of the Japanese Society of Agriculture Machinery), pp. 327–328.

46 Ōmori, Sadao and Hironoshin Takao (1991), 'Hihakai kani hinshitsu hyōka sōchi no kaihatsu (dai 2 hō) (Development of simplified nondestructive quality assessment device (part 2))', *Dai 50 kai Nōgyō Kikai Gakkai nenji taikai kōen yōshi* (Abstracts of papers of the 50th annual conference of the Japanese Society of Agriculture Machinery), p. 395.

47 Ōmori, Sadao and Hironoshin Takao (1992), 'Hihakai kani hinshitsu hyōka sōchi no kaihatsu (dai 4 hō) (Development of simplified nondestructive quality assessment device (part 4))', *Dai 51 kai Nōgyō Kikai Gakkai nenji taikai kōen yōshi* (Abstracts of papers of the 51st annual conference of the Japanese Society of Agriculture Machinery), p. 281.

48 Ōmori, Sadao and Akira Hirata (1993), 'Hikari tōkahō ni yoru kahangata naibu hinshitsu hantei sōchī no kaihatsu (Development of light transmission method-based portable internal quality testing device)', *Dai 52 kai Nōgyō Kikai Gakkai nenji taikai kōen yōshi* (Abstracts of papers of the 52nd annual conference of the Japanese Society of Agriculture Machinery), p. 423.

49 Hirata, Akira and Yōichi Nakamoto (1994), 'Hikari tōkahō ni yoru kahangata naibu hinshitsu hantei sōchī no kaihatsu (dai 2 hō) (Development of light transmission method-based portable internal quality testing device (part 2))', *Dai 53 kai Nōgyō Kikai Gakkai nenji taikai kōen yōshi* (Abstracts of papers of the 53rd annual conference of the Japanese Society of Agriculture Machinery), p. 213.

50 Kondō, Naoshi et al. (1999), 'Ringiku no hinshitsu hyōka ni kansuru kenkyū (dai 1 pō) (Research on quality evaluation of chrysanthemum cut flower (part 1))', *Shokubutsu Kōjō Gakkaishi* (Journal of the Society of High Technology in Agriculture), 11(2): 93–99.

51 Kondō, Naoshi et al. (1999), 'Ringiku no hinshitsu hyōka ni kansuru kenkyū (dai 2 hō) (Research on quality evaluation of chrysanthemum cut flower (part 2))', *Shokubutsu Kōjō Gakkaishi* (Journal of the Society of High Technology in Agriculture), 11(2): 100–105.

52 Kai, Kazuhiro et al. (1995), 'Supurēgiku no kabō fōmēshon no hyōka arugorizumu ni kansuru kenkyū (dai 1 pō) (Research on an algorithm for evaluating spray formation of cut chrysanthemum (part 1))', *Seibutsu kankyō chōsetsu* (Environment control in biology), 33(4): 253–259.

53 Kai, Kazuhiro et al. (1995), 'Supurēgiku no kabō fōmēshon no hyōka arugorizumu ni kansuru kenkyū (dai 2 hō) (Research on an algorithm for evaluating spray formation of cut chrysanthemum (part 2))', *Seibutsu kankyō chōsetsu* (Environment control in biology), 33(4): 261–267.

54 Kai, Kazuhiro et al. (1996), 'Supurēgiku no kabō fōmēshon no hyōka arugorizumu ni kansuru kenkyū (dai 3 pō) (Research on an algorithm for evaluating spray formation of cut chrysanthemum (part 3))', *Seibutsu kankyō chōsetsu* (Environment control in biology), 34(2): 123–128.

55 Hwang, H. et al. (1997), 'Hybrid image processing for robust extraction of lean tissue on beef cut surfaces', *Computers and Electronics in Agriculture*, 17, pp. 281–294.

56 Hwang, H. (2000), 'Extraction of the lean tissue boundary of a beef carcass', *Proceedings of IFAC Bio-Robotics II International Workshop*, pp. 262–265.

57 Daley, W. D. R. et al. (1995), 'Image feature post-processing to characterize visual defects on food products', *Food Processing Automation IV*, pp. 67–77.

58 Park, E. et al. (2005), 'Dynamic thresholding method for improving contaminant detection accuracy with hyperspectral images', *ASAE Paper 053071,* St. Joseph, Michigan: American Society of Agricultural and Biological Engineers.

59 Hwang, H. and S. C. Kim (1999), 'Development of on-line grading system using two surface images of dried oak mushrooms', *Journal of KSAM*, 24(2): 153–158.

60 Hwang, H. (2001), 'Development of automatic grading and sorting system for dry oak mushrooms – 2nd prototype', *Journal of KSAM*, 26(2): 147–154.

61 Ling, P. P. and S. W. Searcy (1991), 'Feature extraction for machine-vision-based shrimp deheader', *Transactions of ASAE*, 34(6): 2631–2636.

62 Little, N. E. at al. (1992), 'Oyster orientation sytem combining computer vision and mechanical technology', *ASAE Paper No. 923514*, St. Joseph, Michigan: American Society of Agricultural and Biological Engineers, pp. 1–26.

63 Nishiokoppe mura website (as at 1 November 2005), http://www.vill.nishiokoppe. hokkaido.jp/.

64 Kawase, K. et al. (2002), 'Terahertz wave parametric source', *Journal of Physics D: Applied Physics*, 35, pp. R1–R14.

65 Wu, Q. et al. (1996), 'Two-dimensional electro-optic imaging of THz beams', *Applied Physics Letters*, 69, p. 1026.

66 Hu, B. B. and M. C. Nuss (1995), 'Imaging with terahertz wave', *Optics Letters*, 1(20): 1716.

67 Watanabe, Y. et al. (2003), 'Component spatial pattern analysis of chemicals using terahertz spectroscopic imaging', *Applied Physics Letters*, 83(4): 800–802.

68 Brucherseifer, M. et al. (2000), 'Label-free probing of the binding state of DNA by time-domain terahertz sensing', *Applied Physics Letters*, 77(24): 4,049–4,051.

69 Endō, T. et al. (2001), 'Spatial estimation of biochemical parameters of leaves with hyperspectral imager', *The 22nd Asian Conference on Remote Sensing*.

70 Endō, T. et al. (2000), 'Estimating net photosynthetic rate based on in-situ hyperspectral data', *Proceedings of SPIE*, 4151, pp. 214–221.

71 Watanabe, Y. et al. (2003), 'Component spatial pattern analysis of chemicals using terahertz spectroscopic imaging', *Applied Physics Letters*, 83(4): 800–802.

Chapter 3

1 Ting, K. C. et al. (1990), 'Robot workcell for seedling transplanting part I – layout and materials flow', *Transactions of the ASAE*, 33(3): 1005–1010.

2 Ting, K. C. et al. (1990), 'Robot workcell for seedling transplanting part II – end-effector development', *Transactions of the ASAE*, 33(3): 1013–1017.

3 Tai, Y. W. et al. (1994), 'Machine vision assisted robotic seedling transplanting', *Transactions of the ASAE*, 37(2): 661–667.

4 Visser International Trade & Engineering B. V. website (as at 1 November 2005), http://www.visserite.com/.

5 Kondō, Naoshi et al. (1999), 'Kiku no sashiki robotto no tameno sashiho bunri kyōkyū shisutemu no kaihatsu (Development of the chrysanthemum cuttings feeding system for the cutting sticking robot)', *Nōgyō Kikai Gakkaishi* (Journal of the Japanese Society of Agriculture Machinery), 61(5): 109–116.

6 Monta, Mitsuji et al. (1998), 'Kiku no sashiki sagyō no jidōka ni kansuru kisoteki kenkyū (dai 3 pō) (Basic studies on automation of chrysanthemum cutting sticking operation (part 3))', *Nōgyō Kikai Gakkaishi* (Journal of the Japanese Society of Agriculture Machinery), 60(4): 37–44.

7 Monta, Mitsuji et al. (1998), 'Kiku no sashiki sagyō no jidōka ni kansuru kisoteki

kenkyū (dai 4 pō) (Basic studies on automation of chrysanthemum cutting sticking operation (part 4))', *Nōgyō Kikai Gakkaishi* (Journal of the Japanese Society of Agriculture Machinery), 60(5): 37–43.

8 Kondō, Naoshi et al. (1998), 'Kiku no sashiki sagyō no jidōka ni kansuru kisoteki kenkyū (dai 1 pō) (Basic studies on automation of chrysanthemum cutting sticking operation (part 1))', *Nōgyō Kikai Gakkaishi* (Journal of the Japanese Society of Agriculture Machinery), 60(2): 67–74.

9 Kondō, Naoshi et al. (1998), 'Kiku no sashiki sagyō no jidōka ni kansuru kisoteki kenkyū (dai 2 hō) (Basic studies on automation of chrysanthemum cutting sticking operation (part 2))', *Nōgyō Kikai Gakkaishi* (Journal of the Japanese Society of Agriculture Machinery), 60(3): 63–70.

10 Arima, Seiichi (1998), 'Tsugiki robotto "Tsuginae Komachi" (Grafting robot "Tsugiki Komachi")', *Kyūshū Baiotekunorojī Kenkyūkai 'haiteku shinpojiumu' kōen yōshi* (Abstracts of papers of Kyūshū Biotechnology Association 'high-tech symposium'), pp. 37–44.

11 Yamada, Hisaya (1996), 'Tsugiki robotto to faitotekunorojī (Grafting robot in phytotechnology), *Nōgyō Kikai Gakkaishi* (Journal of the Japanese Society of Agriculture Machinery), 58(6): 154–158.

12 Gotō, Junichi et al. (1987), 'Edauchi kikai no kikō ni kansuru kenkyū (II) (Basic research on the mechanism of pruning machine (II))', *Nichirinron* (Japanese Forestry Society journal), 98, pp. 713–714.

13 Kondō, Naoshi et al. (1993), 'Budō kanri shūkaku yō robotto no kisoteki kenkyū (dai 1 pō) (Basic studies on robots to work in vineyards (part 1))', *Nōgyō Kikai Gakkaishi* (Journal of the Japanese Society of Agriculture Machinery), 55(6): 85–94.

14 Monta, M. et al. (1995), 'End-effectors for agricultural robot to work in vineyard', *Acta Horticulturae*, 399, pp. 247–254.

15 Nakamura, Yukio et al. (1981), *Nōgyō gijutsu taikei kaju hen 2 budō* (Agricultural technical systems: Fruit tree, vol. 2 Grapes), Tokyo: Nōsangyoson Bunka Kyōkai (Rural Culture Association of Japan).

16 Takatsuji, Masamoto (1989), *Shokubutsu seisan shisutemu jitsuyō jiten* (Practical encyclopedia of bioproduction systems), Tokyo: Fuji Technology Press, pp. 313–315.

17 Monta, Mitsuji et al. (1994), 'Budō kanri shūkaku robotto no kenkyū (dai 3 pō) (Basic studies on robots to work in vineyards (part 3))', *Nōgyō Kikai Gakkaishi* (Journal of the Japanese Society of Agriculture Machinery), 56(2): 93–100.

18 Hayashi, Masahiko et al. (1988), 'Nōyō robotto no kenkyū (Research on agricultural robots), *Dai 6 kai Nihon robotto gakkai gakujutsu kōenkai yōshi* (Abstracts of papers of the 6th annual conference of the Robotics Society of Japan), pp. 579–580.

19 Harrell, R. C. et al. (1990), 'The Florida robotic grove-lab', *Transactions of the ASAE*, 33(2): 391–399.

20 Bourely, A. G. et al. (1990), 'Fruit harvest robotization', *Proceedings of the Agricultural Engineering 1990 Conference of the European Society of Agricultural Engineers.*

21 d'Esnon, A. G. (1985), 'Robotic harvesting of apples', *Agri-Mation*, 1: ASAE, pp. 210–214.

22 Kondo, N. and K. C. Ting (1998), 'Robotics for bioproduction systems', *ASAE*, pp. 231–251.

23 Satō, Tsuneaki et al. (1996), 'Ichigo shūkaku robotto no kaihatsu (Development of strawberry harvesting robot)', *Dai 55 kai Nōgyō Kikai Gakkai nenji taikai kōen yōshi* (Summary of the 55th annual conference of the Japanese Society of Agriculture Machinery), pp. 243–244.

24 Kondō, Naoshi et al. (2001), 'Uchinari saibaiyō ichigo shūkaku robotto (dai 2 hō) (Harvesting robot for strawberries grown on hill top culture (part 2))', *Shokubutsu kōjō gakkaishi* (Journal of the Society of High Technology in Agriculture), 13(4): 231–236.

25 Arima, Seiichi et al. (2004), 'Uchinari saibaiyō ichigo shūkaku robotto no kaihatsu to shūkaku kiso jikken – shūkaku robotto to torēsabiritī shisutemu (Development of strawberry harvesting robot for ridge top culture and basic harvesting experiment)', *Nōgyō kikai gakkai kansai shibu hō* (Journal of the Japanese Society of Agriculture Machinery Kansai Branch), 95, pp. 78–81.

26 Arima, Seiichi et al. (2003), 'Kōsetsu saibaiyō ichigo shūkaku robotto (dai 2 hō) (harvesting robot for Strawberries grown on table top culture (part 2))', *Shokubutsu kōjō gakkaishi* (Journal of the Society of High Technology in Agriculture), 15(3): 162–168.

27 Kondo, N. et al. (1996), 'Visual feedback guided robotic cherry tomato harvesting', *Transactions of the ASAE*, 39(6): 2331–2338.

28 Seiken Sentā (BRAIN) (2005), 'Kasairui shūkaku robotto no kaihatsu (Development of harvesting robots for fruits and vegetables)', *Heisei 16 nendo jigyō hōkoku* (Annual report for 2004), pp. 122–123.

29 Monta, M. et al. (1998), 'End-effectors for tomato harvesting robot', *Artificial Intelligence for Biology and Agriculture*, Dordrecht: Kluwer Academic Publishers, pp. 11–25.

30 Monta, M. et al. (2004), 'Tele-robotics in agriculture – tomato harvesting experiment', *Proceedings of The Second International Symposium on Machinery and Mechatronics for Agriculture and Bio-systems Engineering*, CD-ROM.

31 Monta, M. et al. (1998), 'End-effector for inverted single truss tomato production systems', *Journal of the Japanese Society of Agriculture Machinery*, 60(6): 97–104.

32 Subrata, I. D. M. et al. (1998), 'Sanjigen shikaku sensā wo mochiita minitomato shūkaku robotto (dai 2 pō) (Cherry tomato harvesting robot using 3-D vision sensor

(part 2))', *Nōgyō Kikai Gakkaishi* (Journal of the Japanese Society of Agriculture Machinery), 60(1): 59–68.

33 Han, L. et al. (2000), 'Endoeffekutā ni sanjigen shikaku sensā wo motsu minitomato shūkaku robotto (dai 1 pō) (Cherry tomato harvesting robot with a 3-D vision sensor on its end-effector (part 1))', *Nōgyō Kikai Gakkaishi* (Journal of the Japanese Society of Agriculture Machinery), 62(2): 118–126.

34 Hayashi, Shigehiko et al. (2003), 'V-jigata seishisareta nasu wo taishō to shita robotto shūkaku shisutemu (dai 1 pō) (Robot harvesting system for eggplants trained on a V-shape trellis (part 1))', *Shokubutsu Kōjō Gakkaishi* (Journal of the Society of High Technology in Agriculture), 15(4): 205–210.

35 Hayashi, Shigehiko et al. (2003), 'V-jigata seishisareta nasu wo taishō to shita robotto shūkaku shisutemu (dai 2 hō) (Robot harvesting system for eggplant trained on a V-shape trellis (part 2))', *Shokubutsu Kōjō Gakkaishi* (Journal of the Society of High Technology in Agriculture), 15(4): 211–216.

36 Arima, Seiichi and Naoshi Kondo (1999), 'Cucumber harvesting robot and plant training system', *Journal of Robotics and Mechatronics*, 11(3): 208–212.

37 Van Henten, E. J. et al., (2003), 'Field test of an autonomous cucumber picking robot', *Biosystems Engineering*, 86(3): 305–313.

38 Chung, S. H. et al. (2000), 'Robotto ni yoru kekkyū yasai no sentaku shūkaku no kenkyū (dai 4 hō) (Selective harvesting robot for crisp head vegetables (part 4))', *Nōgyō Kikai Gakkaishi* (Journal of the Japanese Society of Agriculture Machinery), 62(2): 111–117.

39 Murakami, Noriyuki et al. (1999), 'Kyabetsu shūkaku robotto no kaihatsu (dai 1 pō) (Development of robotic cabbage harvester (part 1))', *Nōgyō Kikai Gakkaishi* (Journal of the Japanese Society of Agriculture Machinery), 61(5): 85–92.

40 Murakami, Noriyuki et al. (1999), 'Kyabetsu shūkaku robotto no kaihatsu (dai 2 hō) (Development of robotic cabbage harvester (part 2))', *Nōgyō Kikai Gakkaishi* (Journal of the Japanese Society of Agriculture Machinery), 61(5): 93–100.

41 Sakai, Satoru et al. (2003), 'Nōgyō robotto no tame no jūryōbutsu handoringu manipyurēta (dai 1 pō) (A heavy material handling manipulator for agricultural robot (part 1))', *Nōgyō Kikai Gakkaishi* (Journal of the Japanese Society of Agriculture Machinery), 65(4): 108–116.

42 Sakai, Satoru et al. (2003), 'Nōgyō robotto no tame no jūryōbutsu handoringu manipyurēta (dai 2 hō) (A heavy material handling manipulator for agricultural robot (part 2))', *Nōgyō Kikai Gakkaishi* (Journal of the Japanese Society of Agriculture Machinery), 65(4): 117–123.

43 Sakai, Satoru et al. (2003), 'Control of a heavy material handling agricultural manipulator using robust gain-scheduling and μ-synthesis', *Proceedings of IEEE International Conference on Robotics and Automation*, pp. 96–102.

44 Sakai, Satoru and Mikio Umeda (2002), 'Robasuto seigyo riron wo mochiita

nōgyō robotto no Minimal Equipment (Minimal Equipment for agricultural robot using robust control theory)', *Dai 61 kai Nōgyō Kikai Gakkai nenji taikai kōen yōshi* (Abstracts of papers of the 61st annual conference of the Japanese Society of Agriculture Machinery), pp. 303–304.

45 Hwang, H. and S. C. Kim (2000), 'Tele-operative system for bio-production – remote local image processing for object identification', *Proceedings of Bio-Robotics II International Workshop*, pp. 188–191.

46 Kim, S. C. and H. Hwang (2003), 'Tele-operative task environment interface with identification of greenhouse watermelon under natural light', *Proceedings of IEEE/ASME International Conference on Advanced Intelligent Mechatronics*, pp. 1,250–1,354.

47 Hwang, H. and S. C. Kim (2003), 'Development of multi-functional tele-operative modular robotic system for greenhouse watermelon', *Proceedings of IEEE/ASME International Conference on Advanced Intelligent Mechatronics*, pp. 1344–1349.

48 BRAIN (Seibutsukei Tokutei Sangyō Gijutsu Kenkyū Suishin Kikō) (1997), *Heisei 8 nendo jigyō hōkoku* (Annual report for 1996), pp. 12–15.

49 Aikoku Alpha Corporation RH Division website (as at 1 November 2005): http://222. aikoku.co.jp/rh/.

50 Toyo Koken K. K. website (as at 1 November 2005), http://www.toyokoken.co.jp.

51 Reed, J. N. et al. (2001), 'Automatic mushroom harvester development', *Journal of Agricultural Engineering Research*, 78(1): 15–23.

52 Ishii, Tōru et al. (2003), 'Rakuyōkei kajitsu senbetsu robotto (dai 1 pō) (Deciduous fruit grading robot (part 1))', *Nōgyō Kikai Gakkaishi* (Journal of the Japanese Society of Agriculture Machinery), 65(6): 163–172.

53 Ishii, Tōru et al. (2003), 'Rakuyōkei kajitsu senbetsu robotto (dai 2 hō) (Deciduous fruit grading robot (part 2))', *Nōgyō Kikai Gakkaishi* (Journal of the Japanese Society of Agriculture Machinery), 65(6): 173–183.

54 Omi Weighing Machine Inc. website (as at 1 November 2005), http://www.omi-scale. com/index2.html.

55 Australian Wool Corporation (1988), 'Shearing with robots', *Wool Research and Development, Highlights of the 1987–1988 Program*, Sydney, pp. 22–25.

56 Australian Wool Corporation (1989), 'Robot shearing update', *Wool Research and Development, Highlights of the 1988–1989 Program*, Sydney, pp. 36–37.

57 Trevelyan, J. P. (1992), *Robots for Shearing Sheep: Shear Magic*, London: Oxford University Press.

58 Notsuki, Iwao (1979), 'Zenjidō sakunyūki (Fully automatic milking machine)', *Nōrin suisan gijutsu kaigi jimukyoku kenkyū seika No. 144: sakunyū sagyō no shōryokuka ni kansuru kenkyū* (AFFRC Secretariat research result No. 144: Research on labor saving in milking operation), pp. 156–182.

59 Ichito, Kazutomo (1998), 'Sakunyū robotto kaihatsu kenkyū no genjō to tenbō (The

current status and outlook of research and development of milking robots)', *Nōgyō Kikai Gakkaishi* (Journal of the Japanese Society of Agriculture Machinery), 60(6): 138–142.

60 Honda, Yoshifumi (2001), 'Sakunyū sagyō no robottoka (Robotization of milking operation)', *Nōrinsuisan gijutsu kenkyū jānaru* (Research journal of food and agriculture), 24(6): 38–45.

61 Hirata, Akira et al. (1998), 'Tsunagikai yō sakunyū robotto no kaihatsu kenkyū (dai 1 pō) (Research and development of milking robot for stanchion stall (part 1))', *Dai 57 kai Nōki Gakkai Kōen Yōshi* (Abstracts of papers of the 57th conference of the Japanese Society of Agriculture Machinery), pp. 425–426.

62 *Gripping Apparatus with Two Fingers Covered by a Moveable Film*, International Patent Application WO03011536, UK Patent Application GB2378432.

63 Reed, J. N. and S. J. Miles (2004), 'High-speed conveyor junction based on an air-jet floatation technique', *Mechatronics*, 14(6): 685-699.

Chapter 4

1 BRAIN (Seibutsukei Tokutei Sangyō Gijutsu Kenkyū Suishin Kikō) (2002), *Heisei 13 nendo jigyō hōkoku* (Annual report for 2001), pp. 10–11.

2 BRAIN (Seibutsukei Tokutei Sangyō Gijutsu Kenkyū Suishin Kikō) (2003), *Heisei 14 nendo jigyō hōkoku* (Annual report for 2002), pp. 10–11.

3 Hamada, Yasuyuki et al. (2003), 'Unten shien yō sagyō nabigēta (Operation navigator for driving support)', *Dai 62 kai Nōgyō Kikai Gakkai nenji taikai kōen yōshi* (Abstracts of papers of the 62nd annual conference of the Japanese Society of Agriculture Machinery), pp. 475–476.

4 Deere & Company catalogue.

5 Yukumoto, Osamu et al. (2001), 'Kōun sagyō wo okonau jiritsu idō robotto ni kansuru kenkyū (Research on autonomous guidance robot for tillage operation)', *Nōgyō Kikaika Kenkyūjo Hōkoku* (Technical report of the Institute of Agricultural Machinery), 32, pp. 1–96.

6 Masuda, Yūichi (2002), 'Jiritsu chokushin sōchi no kaihatsu (Development of autonomous straight traveling device)', *Nōgyō Kikai Gakkaishi* (Journal of the Japanese Society of Agriculture Machinery), 64(5): 30–31.

7 Matsuo, Yōsuke (2003), 'Jidō chokushin taueki no kaihatsu (Development of auto-steering rice transplanter)', *BRAIN tekuno news* (BRAIN techno news), 99, pp. 36–39.

8 Inoue, Keiichi et al. (1999), 'Jiritsu sōkō no tame no GPS to jairo no karuman firutā ni yoru sensa fūjon gijutsu (dai 1 pō) (Sensor fusion technology for GPS and gyro by Kalman filter for automatic guidance (part 1))', *Nōgyō Kikai Gakkaishi* (Journal of the Japanese Society of Agriculture Machinery), 61(4): 102–113.

9 Matsuo, Yōsuke et al. (1995), 'Bijon to chijiki hōi sensā wo mochiita jiritsu sōkō no

kenkyū (Research on automatic guidance using vision and geomagnetic direction sensor)', *Heisei 2–5 nendo jutaku kenkyū hōkokusho* (Funded research report 1990–1993): Seibutsukei Tokutei Sangyō Gijutsu Kenkyū Suishin Kikō, pp. 1–44.

10 Tillett, N. D., T. Hague and J. A. Marchant (1998), 'A robotic system for plant-scale husbandry', *Journal of Agricultural Engineering Research*, 69(2): 169–178.

11 Tian, L. et al. (1999), 'Development of a precise sprayer for site-specific weed management', *Transactions of the ASAE*, 42(4): 893–900.

12 Kise, M. et al. (2005), 'A stereovision-based crop row detection method for tractor-automated guidance', *Biosystems Engineering*, 90(5): 357–367.

13 Yukumoto, Osamu et al. (2000), 'Jidō tsuijū gijutsu ni kansuru kenkyū – jisō sharyō ni yoru jidō tsuijū (Research on automatic following technology – automatic following by automatic guidance vehicle)', *Shonai tokuken (ōgata) heisei 7–11 nendo sōkatsu hōkokusho* (General report on inhouse special research (large projects) 1995–1999), Tokyo: Seiken, pp. 59–73.

14 Yukumoto, Osamu et al. (2003), 'Haru kyabetsu wo taishō to shita sentakushiki shūkakuki to jidō tsuijū unpansha no kaihatsu (dai 1 pō) (Development of automatic following transport vehicle and selective harvester for spring cabbages (part 1))', *Nōgyō Kikai Gakkai Kantō Shibu dai 39 kai nenji hōkoku* (The 39th annual report of the Japanese Society of Agriculture Machinery Kantō Branch), pp. 20–21.

15 Sutiarso, L. et al. (2002), 'Trajectory control and its application to approach a target: Part II. Target approach experiments', *Transactions of the ASAE*, 45(4): 1199–1205.

16 Okado, Atsushi et al. (2000), 'Jiritsu sōkō shūkakuki wo kijiku to shita shūkaku unpan sagyō kyōchō shisutemu no kaihatsu (Development of coordinated harvesting and transportation system centering on automatic guidance harvesting machine)', *Heisei 11 nendo sōgō nōgyō shiken kenkyū seiseki keikaku gaiyōshū* (An overview of agricultural testing and research results and project 1999), Tokyo: National Agriculture Research Centre, pp. 55–56.

17 Iida, Michihisa (2002), 'Konbain no gun kanri shisutemu (Multiple combine control system)', *Nōgyō kikai gakkai shinpojiumu 'mirai no shokuryō seisan wo sasaeru nōgyō robotto jidōka fōramu' kōen ronbunshū* (Collected papers of the JSAM symposium 'Forum on agricultural robots and automation for food production in the future'), Tokyo: Japanese Society of Agriculture Machinery, pp. 35–46.

18 Noguchi, Noboru (2001), 'Sharyōkei nōgyō kikai no robottoka (Robotization of vehicle type agricultural machinery)', *Nōrinsuisan gijutsu kenkyū jānaru* (Research journal of food and agriculture), 24(6): 5–10.

19 Kise, Michio et al. (2001), 'RTK-GPS to FOG wo shiyōshita hojō sagyō robotto (dai 1 pō) (Field operation robot using RTK-GPS and FOG (part 1))', *Nōgyō Kikai Gakkaishi* (Journal of the Japanese Society of Agriculture Machinery), 63(5): 74–79.

20 Yukumoto, Osamu et al. (1997), 'Kōun robotto no kōhō gijutsu (Navigation technology of tilling robot)', *Nōgyō Kikai Gakkai Hokkaido Shibu kokusai shinpojiumu shiryō* (Papers of the JSAM Hokkaido Branch international symposium), Tokyo: Japanese Society for Agriculture Machinery, pp. 79–94.

21 Yukumoto, Osamu et al. (1998), 'Kōun robotto shisutemu no kaihatsu (dai 3 pō) (Development of tilling robots using position sensing system and geomagnetic direction sensor (part 3))', *Nōgyō Kikai Gakkaishi* (Journal of the Japanese Society of Agriculture Machinery), 60(5): 53–61.

22 Matsuo, Yosuke et al. (2001), 'Kōun robotto no kōhō gijutsu to sagyō seinō (dai 1 pō), (dai 2 hō) (Navigation technology and operation performance of tilling robot (part 1), (part 2))', *Nōgyō Kikai Gakkaishi* (Journal of the Japanese Society of Agriculture Machinery), 63(3): 114–129.

23 Nagasaka, Yoshisada (2003), 'Shokuryō seisan wo sasaeru nōgyō robotto (Agricultural robots supporting food production)', *Nihon Kikai Gakkaishi* (Journal of the Japan Society of Mechanical Engineers), 106(1015): 458–459.

24 Yamashita, Mitsushi et al. (1993), 'Jōyō taueki no tame no sōkō seigyo shisutemu no kaihatsu (Development of steering control system for riding type rice transplanter)', *Nōgyō Kikai Gakkai Kansai Shibu hō* (Report of the Japanese Society of Agriculture Machinery Kansai Branch), 74, pp. 47–48.

25 Nonami, Kazuyoshi et al. (1993), 'Jōyō taueki no sōkō seigyo ni kansuru kenkyū (dai 1 pō) (Research on travel control for riding type rice transplanter (part 1))', *Nōgyō Kikai Gakkaishi* (Journal of the Japanese Society of Agriculture Machinery), 55(4): 107–114.

26 Seki, Masahiro et al. (2000), 'Kankyō chōwa gata josō robotto shisutemu no kaihatsu (Development of environmentally friendly weeding robot system (part 3))', *Heisei 11 nendo sōgō nōgyō shiken kenkyū seiseki keikaku gaiyōshū* (An overview of agricultural experiment and research results and projects 1999), Tokyo: National Agricultural Research Center, pp. 13–14.

27 Seki, Masahiro et al. (2000), 'Kankyō chōwa gata josō robotto shisutemu no kaihatsu (dai 3 pō) (Development of environmentally friendly weeding robot system (part 3))', *Dai 59 kai Nōgyō Kikai Gakkai nenji taikai kōen yōshi* (Abstracts of papers of the 59th annual conference of the Japanese Society of Agriculture Machinery), Tokyo: Japanese Society of Agriculture Machinery, pp. 51–52.

28 Yamashita, Jun et al. (1991), 'Shisetsu engei yō mujin unpansha no shisaku to sono jitsuyō seinō (Prototype unmanned transport vehicle for protected horticulture and its practical performance)', *Nōgyō Kikai Gakkaishi* (Journal of the Japanese Society of Agriculture Machinery), 53(5): 75–84.

29 Monta, M. (1998), 'Traveling devices within bioproduction environments', *Robotics for Bioproduction Systems*: ASAE, pp. 133–135.

30 Hayashi, Shigehiko et al. (2003), 'V-jigata seishisareta nasu wo taishō to shita robotto

shūkaku shisutemu (dai 1 pō) (Robot harvesting system for eggplants trained on a V-shape trellis (part 1))', *Shokubutsu kōjō gakkaishi* (Journal of the Society of High Technology in Agriculture), 15(4): 205–210.

31 BRAIN (Seibutsukei Tokutei Sangyō Gijutsu Kenkyū Suishin Kikō) (2002), *Heisei 13 nendo jigyō hōkoku* (Annual report for 2001), pp. 110–111.

32 Maruyama Mfg. Co., Inc. (2000), *Shatoru supurēka MCS5-100U toriatsukai setsumeisho* (Shuttle spray-car MSC5-100U operation manual).

33 KIORITZ Corporation (1990), *Robotto supurēka gijutsu shiryō* (Robot spray-car technical report).

34 Maruyama Mfg. Co., Ltd. (2003), *Shatoru supurēka no katarogu* (Shuttle spray-car catalogue).

35 KIORITZ Corporation (2003), *Robotto supurēka katarogu* (Robot spray-car catalogue).

36 Tosaki, Kōichi et al. (1996), 'Yūdō kēburu shiki kaju mujin bōjoki no kaihatsu (dai 1 pō) (Development of guide cable type unmanned pest control machine for fruit trees (part 1))', *Nōgyō Kikai Gakkaishi* (Journal of the Japanese Society of Agriculture Machinery), 58(6): 101–110.

37 Nōgyō Kikai Gakkai (Japanese Society of Agriculture Machinery) (ed.) (1996), *Seibutsu seisan kikai handobukku* (Handbook of bio-production machines), Tokyo: Corona Publishing, p. 731.

38 Cho, S. I. and J. H. Lee (2000), 'Autonomous speed-sprayer using differential GPS system, genetic algorithm and fuzzy control', *Journal of Agricultural Engineering Research*, 76, pp. 111–119.

39 Ollis, M. and A. Stentz (1997), 'Vision-based perception for an automated harvester', *Proceedings of the IEEE/RSJ International Conference on Intelligent Robotic Systems*.

40 Hoffman, R. M. et al. (1998), 'Demeter: an autonomous alfalfa harvesting system', *ASAE Paper No. 983005*, St. Joseph, Michigan: American Society of Agricultural and Biological Engineers.

41 Kubota, Yōsuke (1998), 'Kusakari robotto jiritsu sōkō shisutemu no gijutsu shōkai (Information on automatic guidance system for mowing robot)', *Keisoku jidō seigyo gakkaishi* (SICE journal), pp. 91–96.

42 Okuyama, S. et al. (1987), 'Automatic control of a lawn tractor', *Proceedings*, Tokyo: International Symposium on Agricultural Mechanization and International Cooperation in High Technology Era, University of Tokyo, pp. 171–178.

43 Deere & Company website (as at 1 November 2005), http://www.deere.com/en_US/deerecom/usa_canada.html.

44 Tamaki, Katsuhiko et al. (2001), 'Keisha kusachi ni okeru hirohaba shihi sagyō ni muketa jidō sōkō seigyo gijutsu (dai 1 pō) (Automatic guidance control technology for broadcast fertilizer application on sloping grassland (part 1))', *Nōgyō Kikai*

Gakkaishi (Journal of the Japanese Society of Agriculture Machinery), 63(5): 109–115.

45 Motohashi, Kuniji et al. (1996), 'Sokukyorin ni yoru ichi ninshiki (Position recognition by measuring wheel), *Nōgyō Kikai Gakkaishi* (Journal of the Japanese Society of Agriculture Machinery), 58(1): 43–48.

46 Kanemitsu, Mikio et al. (2003), 'Keishachi kaju yō tamokuteki monorēru no kaihatsu (dai 2 hō) (Development of multipurpose monorail for fruits growing on sloping orchards (part 2))', *Dai 62 kai Nōgyō Kikai Gakkai nenji taikai kōen yōshi* (Abstracts of papers of the 62nd annual conference of the Japanese Society of Agriculture Machinery), pp. 83–84.

47 Okazaki, Kōichirō et al. (1993), 'Jujō sōkō monorēru yō kajitsu hansō sōchi no kaihatsu (Development of fruit transportation device for monorail traveling above tree canopies)', *Nōgyō Kikai Gakkai Kansai Shibu hō* (Report of the Japanese Society of Agriculture Machinery Kansai Branch), 73(6): 33–36.

48 Nōgyō Kikai Gakkai (ed.) (1988), *Nōgyō kikai no robottoka ni kansuru chōsa kenkyū – chōsa hōkokusho* (Survey research on robotization of agricultural machinery – survey report), Tokyo: Japanese Society of Agriculture Machinery, pp. 105–143.

49 Kanasu, Masayuki and Shigeru Yagi (1976), 'Hojō sagyō sōchi (Field operation equipment)', *Nōgyō Kikai Gakkaishi* (Journal of the Japanese Society of Agriculture Machinery), 38(3): 452–453.

50 Yamashita, Jun et al. (2002), 'Ichigo saibaiyō gantori shisutemu no kaihatsu kenkyū (dai 1 pō) (Research and development of gantry system for strawberry culture (part 1))', *Nōgyō Kikai Gakkaishi* (Journal of the Japanese Society of Agriculture Machinery), 64(2): 122–130.

51 Shin Enerugī Sangyō Gijutsu Sōgō Kaihatsu Kikō et al. (1999), *Daikibo nōgyō muke seimitsu jiritsu sōkō sagyō shien shisutemu kaihatsu kenkyū konsōshiamu hōkokusho (dai 1 nendo)* (Report of the consortium for R & D of precision autonomous traveling work supporting system for large-scale farming (year 1)).

52 Shin Enerugī Sangyō Gijutsu Sōgō Kaihatsu Kikō et al. (2000), *Daikibo nōgyō muke seimitsu jiritsu sōkō sagyō shien shisutemu kaihatsu kenkyū konsōshiamu hōkokusho (dai 2 nendo)* (Report of the consortium for R & D of precision autonomous traveling work supporting system for large-scale farming (year 2)).

53 Shin Enerugī Sangyō Gijutsu Sōgō Kaihatsu Kikō et al. (2001), *Daikibo nōgyō muke seimitsu jiritsu sōkō sagyō shien shisutemu kaihatsu kenkyū konsōshiamu hōkokusho (dai 3 nendo)* (Report of the consortium for R & D of precision autonomous traveling work supporting system for large-scale farming (year 3)).

54 Hirose, Shigeo and Akihiko Nagakubo (1992), 'Hokōgata hekimen idō robotto (Legged wall climbing robot)', *Nihon Robotto Gakkaishi* (Journal of the Robotic Society of Japan), 10(5): 575–580.

55 Nōgyō Kikai Gakkai (ed.) (2004), seminar information 'Anzen de anshin na shakai

ni kōkensuru sangyō yō mujin herikoputā (Industrial unmanned helicopter for a safe and secure society)', Tokyo: Japanese Society of Agriculture Machinery.

56 Noguchi, N. et al. (2004), 'Development of a master-slave robot system for farm operations', *Computers and Electronics in Agriculture*, 44(1): 1–19.

57 Ishii, K. and N. Noguchi (2004), 'Task management and control system for multi-robot using wireless LAN', *Proceedings of ASAE Conference of Automation Technology for Off-Road Equipment*, pp. 56–63.

58 Abe, Tsuyoshi et al. (2005), 'Rēzā sukyana wo mochiita nōyō sharyō no cyokushin tsuijū seigyo ni kansuru kenkyū (Research on straight follow-up traveling control of agricultural vehicles using laser scanner)', *Nōgyō Kikai Gakkaishi* (Journal of the Japanese Society of Agriculture Machinery), 67(3): 65–71.

59 Noguchi, Noboru (203), 'Sangyōyō mujin herikoputa wo tekiyōshita hojō kankyō monitaringu (Field environmental monitoring using industrial unmanned helicopter)', *Nōgyō Kikai Gakkaishi* (Journal of the Japanese Society of Agriculture Machinery), 65(4): 13–17.

60 Sugiura, R. et al. (2005), 'Remote sensing technology for vegetation monitoring using an unmanned helicopter', *Biosystems Engineering*, 90(4): 369–379.

61 Sugiura, Ryō et al. (2005), 'Herikoputa bēsu rimōto senshingu no tame no hikō monitaringu shisutemu (Flight monitoring system for helicopter-based remote sensing)', *Nōgyō Kikai Gakkaishi* (Journal of the Japanese Society of Agriculture Machinery), 67(2): 86–92.

Index